Automated Machine Learning on AWS

Fast-track the development of your production-ready machine learning applications the AWS way

Trenton Potgieter

BIRMINGHAM—MUMBAI

Automated Machine Learning on AWS

Copyright © 2022 Packt Publishing

Publishing Product Manager: Devika Battike
Senior Editor: Nathanya Dias
Content Development Editor: Nazia Shaikh
Technical Editor: Devanshi Ayare
Copy Editor: Safis Editing
Project Coordinator: Aparna Ravikumar Nair
Proofreader: Safis Editing
Indexer: Sejal Dsilva
Production Designer: Roshan Kawale
Marketing Coordinator: Abeer Dawe, Shifa Ansari

First published: April 2022
Production reference: 1100322

Published by Packt Publishing Ltd.
Livery Place
35 Livery Street
Birmingham
B3 2PB, UK.

ISBN 978-1-80181-182-8

www.packt.com

Foreword

Virtually everyone struggles with operationalizing machine learning models. Training your first model can sometimes seem like an insurmountable challenge, until you realize that you also need an end-to-end pipeline to supply new data for inference and retraining the model when its performance inevitably degrades. Although AWS offers the broadest and deepest set of machine learning services, figuring out where to get started and how to tie all those options together normally requires months of painful experimentation. This book cuts through the uncertainty based on **Trenton's** first-hand experiences working with both the most sophisticated technology companies in the world as well as organizations new to machine learning.

I've worked with hundreds of companies around the world trying to get value from artificial intelligence and machine learning. The problem is that machine learning can mean very different things even within the same company, much less across different organizations or industries. Some teams are just starting to invest in AI and machine learning and want to build their first model, while other teams in the same organization want to scale up sophisticated experimentation and monitoring frameworks to support thousands of models in production. Most companies hire data scientists or machine learning engineers with skill mismatches in the hope that they'll figure it out. Trenton has the rare advantage of seeing how large organizations have successfully scaled up their modeling pipelines as well as where they've faltered. Even more importantly, he has hard-won experience helping them solve those challenges.

The machine learning space evolves so quickly that focusing on any single algorithm, package, or platform can lead to outdated content. Trenton avoids this trap by translating timeless software engineering concepts like continuous integration and continuous delivery to the machine learning space. Unlike many approaches, however, he punctuates each concept with hands-on examples to illustrate how everything works in practice so that you don't need to struggle to translate theory to real life applications.

For example, data scientists often view automated machine learning with disdain due to previous exposure to automation that felt more like a straitjacket than an accelerant. People new to machine learning as well as sophisticated data scientist can overlook AutoML on AWS due to inexperience or ignorance of its benefits. Understanding when and why to use AutoML to get an initial benchmark on a new project or avoid manually selecting and tuning algorithms every time you retrain a model can reduce the time you spend on model training by an order of magnitude.

Even more importantly, learning how to think about the long-term maintenance of the machine learning pipelines will help you avoid painful decisions on whether to spend time refactoring existing models or deliver new projects. Software engineers have been leveraging CI/CD processes for over a decade at this point, but most machine learning practitioners aren't aware of best practices from the DevOps space. Most data scientists discover the need for this process only after they've built a few models and realized that reusable model assets and pipelines are required if they want to do anything beyond maintaining brittle modeling workflows by hand.

Finally, Trenton highlights concepts like source-code and data-centric machine learning that normally require hiring working at a top technology company that's overcome scaling challenges that most companies don't experience early on in their machine learning journeys. Most people and organizations hit a wall after implanting a CI/CD pipeline and building their first. They run up against the challenges of scheduling, tracking, and monitoring their machine learning pipelines. This book is the only example I'm aware of that offers prescriptive guidance on how to structure long-term machine learning pipelines and avoid the common pitfalls that machine learning teams typically encounter.

In short, the concepts in this book will help you move beyond the hopes and dreams of machine learning, to getting machine learning applications into production and delivering value.

Jonathan Dahlberg

Head of ML Solution Engineering

Snorkel AI

Contributors

About the author

Trenton Potgieter is a senior AI/ML specialist at AWS and has been working in the field of ML since 2011. At AWS, he assists multiple AWS customers to create ML solutions and has contributed to various use cases, broadly spanning computer vision, knowledge graphs, and ML automation using MLOps methodologies. Trenton plays a key role in evangelizing the AWS ML services and shares best practices through forums such as AWS blogs, whitepapers, reference architectures, and public-speaking events. He has also actively been involved in leading, developing, and supporting an internal AWS community of MLOps-related subject matter experts.

About the reviewer

Hemanth Boinpally is a Machine Learning Engineer at AWS. He has several years of experience working in data science and ML. He has worked with enterprise customers across different industries, such as healthcare, finance, logistics, and manufacturing. He enjoys providing end-to-end ML solutions for complex business problems. His expertise across the technology stack helps him collaborate with cross-functional teams to build successful ML products. This includes engaging with business stakeholders, the research and development of ML models, and operationalizing these models using MLOps principles. He has worked in areas such as model bias detection, interpretable models, NLP, CV, active learning, and deep learning.

Table of Contents

3
Automating Complicated Model Development with AutoGluon

Section 2: Automating the Machine Learning Process with Continuous Integration and Continuous Delivery (CI/CD)

4
Continuous Integration and Continuous Delivery (CI/CD) for Machine Learning

5

Continuous Deployment of a Production ML Model

Section 3: Optimizing a Source Code-Centric Approach to Automated Machine Learning

6

Automating the Machine Learning Process Using AWS Step Functions

7
Building the ML Workflow Using AWS Step Functions

Section 4: Optimizing a Data-Centric Approach to Automated Machine Learning

8
Automating the Machine Learning Process Using Apache Airflow

9
Building the ML Workflow Using Amazon Managed Workflows for Apache Airflow

Section 5: Automating the End-to-End Production Application on AWS

10
An Introduction to the Machine Learning Software Development Life Cycle (MLSDLC)

11
Continuous Integration, Deployment, and Training for the MLSDLC

Preface

AWS provides a wide range of solutions to help automate a **machine learning** (**ML**) workflow with just a few lines of code. With this practical book, you'll learn how to automate an ML pipeline using the various AWS services.

Automated Machine Learning on AWS begins with a quick overview of what the ML pipeline/process looks like and highlights the typical challenges you may face when building a pipeline. By reading the book, you'll become well versed in various AWS solutions, such as Amazon SageMaker Autopilot, AutoGluon, AWS Step Functions, and more, and will learn how to automate an end-to-end ML process with the help of hands-on examples. The book will show you how to build, monitor, and execute a CI/CD pipeline for the ML process and how the various CI/CD services within AWS can be applied to a use case with the **Cloud Development Kit** (**CDK**). You'll understand what a data-centric ML process is by working with Amazon Managed Services for Apache Airflow and will build a managed Airflow environment. You'll also cover the key success criteria for an **Machine Learning Software Development Life Cycle** (**MLSDLC**) implementation and the process of creating a self-mutating CI/CD pipeline using the CDK from the perspective of the platform engineering team.

By the end of the book, you'll be able to effectively automate a complete ML pipeline and deploy it to production.

Who this book is for

This book is for novice as well as experienced ML practitioners looking to automate the process of building, training, and deploying ML-based solutions into production, using both purpose-built and other AWS services. A basic understanding of the end-to-end ML process and concepts, Python programming, and AWS is necessary to make the most out of the book.

What this book covers

Chapter 1, Getting Started with Automated Machine Learning on AWS, provides an overview of what the ML pipeline/process looks like and highlights the typical challenges you will face when building the pipeline. The main challenge to highlight is overcoming the interactive nature of the process and why automation is crucial to a successful process. Subsequently, we will introduce the concept of AutoML and highlight how it can alleviate the aforementioned challenges.

Chapter 2, Automating Machine Learning Model Development Using SageMaker Autopilot, provides an overview of what SageMaker Autopilot is and how it can be useful in automating the ML process. By using an example use case (ACME Fishing Logistics), the chapter will further educate you on how to practically leverage SageMaker Autopilot and apply it to the use case. The chapter accomplishes this by walking you through each step of the process, comparing it to the model framing example to highlight the benefits of process automation.

Chapter 3, Automating Complicated Model Development with AutoGluon, provides you with an overview of what AutoGluon is, how it differs from SageMaker Autopilot, and the value it adds for use cases that involve deep learning models that make use of text, image, and tabular data. It further elaborates on AutoGluon's capabilities for process automation by walking you through the hands-on, ACME Fishing Logistics example, and a deep learning-based model for computer vision.

Chapter 4, Continuous Integration and Continuous Delivery (CI/CD) for Machine Learning, introduces you to the concept of **continuous integration and continuous deployment (CI/CD)** and how specifically it can be applied to an ML use case. The chapter accomplishes this by introducing DevOps culture and highlighting how the DevOps process can evolve into an MLOps process. This chapter also introduces and focuses on how the various CI/CD services within AWS can be applied to the use case, by introducing you to the **Cloud Development Kit (CDK)** and the Cloud9 development environment. The chapter will also practically show you how to set up the development workspace, install and configure the CDK, set up the artifact repositories, and start codifying the primary artifacts that will be leveraged by the CI/CD pipeline.

Chapter 5, Continuous Deployment of a Production ML Model, introduces you to the typical tasks performed by the ML practitioner, within the context of the deployed CI/CD pipeline and DevOps culture. The chapter will walk you through creating the model assets, which trigger the pipeline execution, and show you how to manage and monitor the progress.

Chapter 6, Automating the Machine Learning Process Using AWS Step Functions, highlights how the CI/CD process can be further optimized, by including the ML practitioner in the majority of the pipeline build process. This chapter shows how this can be done by introducing AWS Step Functions and the Data Science SDK for Step Functions. It will then walk you through how to integrate the Data Science SDK into the CI/CD pipeline process.

Chapter 7, Building the ML Workflow Using AWS Step Functions, elaborates on the role and tasks of the ML practitioner, within the context of further optimizing the CI/CD pipeline, by walking you through how to build the codified ML workflow, perform integration testing on the workflow, and deploy the ML model into production, using the workflow.

Chapter 8, Automating the Machine Learning Process Using Apache Airflow, introduces you to a data-centric workflow, why its application to the ML process is important, and the team members normally responsible for executing this part of the process. The chapter elaborates on the common tools used to perform this function, namely Apache Airflow, and the Amazon managed service for Apache Airflow. The chapter will then walk you through how to build a managed Airflow environment.

Chapter 9, Building the ML Workflow Using Amazon Managed Workflows for Apache Airflow, leverages the environment created in the previous chapter and focuses on the role and tasks that the ML practitioner performs, within the context of further optimizing the CI/CD pipeline. The chapter accomplishes this by walking you through how to build the codified ML workflow, perform integration testing on the workflow, and deploy the ML model into production, using the workflow running on the MWAA environment.

Chapter 10, An Introduction to the Machine Learning Software Development Life Cycle (MLSDLC), introduces you to the MLSDLC methodology and explains why adopting this methodology encompasses a holistic solution for automating the entirety of the ML-based application. The chapter highlights the key success criteria for an MLSDLC implementation – the cross-functional and agile team. It showcases this success criteria by walking through each of the team member roles, how they interact with the other team members, and building the codified artifacts that each role is responsible for.

Chapter 11, Continuous Integration, Deployment, and Training for the MLSDLC, walks through the process of creating a self-mutating CI/CD pipeline using the CDK, from the perspective of the platform engineering team. The chapter will show you how to take the various cross-functional teams' artifacts and combine them into an automated process for CI of both the ACME Fishing Logistics application and the ML model in a development and QA environment. The chapter will also highlight how to include automated integration and QA test procedures for the web application, plus ML model inferences, in the overall MLSDLC workflow. The chapter will then show you how to take the application from the test environment into the production environment, to produce the production version of the overall ML application. The last part of the chapter will focus on the various tasks and procedures from the perspective of the data engineering team, to essentially *close the loop* on the MLSDLC process, by walking you through how to apply continuous training of the pipeline, based on new data and the lessons learned from *chapter 8, Automating the Machine Learning Process Using Apache Airflow*.

To get the most out of this book

You will need a functional AWS account to run the examples.

Software/hardware covered in the book	Operating system requirements
Python 3.7.10 (and above)	Windows/macOS/Linux
AWS CLI 1.19.112 (and above)	Windows/macOS/Linux
AWS CDK 2.3.0 (build beaa5b2)	Windows/macOS/Linux

It is recommended that you use an AWS Cloud9 integrated development environment as it meets the software/hardware and operating system requirements.

If you are using the digital version of this book, we advise you to type the code yourself or access the code from the book's GitHub repository (a link is available in the next section). Doing so will help you avoid any potential errors related to the copying and pasting of code.

Where possible, applicable AWS services have been used to automate the example ML workflow. We encourage you to review how the provided examples could be further adapted to use additional AWS services, such as Amazon SageMaker Pipelines, or even open source alternatives, such as Kubeflow Pipelines.

Download the example code files

You can download the example code files for this book from GitHub at `https://github.com/PacktPublishing/Automated-Machine-Learning-on-AWS`. If there's an update to the code, it will be updated in the GitHub repository.

We also have other code bundles from our rich catalog of books and videos available at `https://github.com/PacktPublishing/`. Check them out!

Download the color images

We also provide a PDF file that has color images of the screenshots and diagrams used in this book. You can download it here: `https://static.packt-cdn.com/downloads/9781801811828_ColorImages.pdf`.

Conventions used

There are a number of text conventions used throughout this book.

`Code in text`: Indicates code words in text, database table names, folder names, filenames, file extensions, pathnames, dummy URLs, user input, and Twitter handles. Here is an example: "We define a `train()` function to capture the input parameters and fit an `ImagePredictor()` to `training_data`."

A block of code is set as follows:

```
import boto3
import sagemaker
aws_region = sagemaker.Session().boto_session.region_name
!sm-docker build --build-arg REGION={aws_region} .
```

When we wish to draw your attention to a particular part of a code block, the relevant lines or items are set in bold:

```
import sagemaker
import datetime
image_uri = "<Enter the Image URI from the sm-docker output>"
role = sagemaker.get_execution_role()
session = sagemaker.session.Session()
bucket = session.default_bucket()
```

Bold: Indicates a new term, an important word, or words that you see onscreen. For instance, words in menus or dialog boxes appear in **bold**. Here is an example: "Using the **Amazon SageMaker** management console, click the **Open SageMaker Studio** button."

> **Tips or important notes**
> Appear like this.

Get in touch

Feedback from our readers is always welcome.

General feedback: If you have questions about any aspect of this book, email us at customercare@packtpub.com and mention the book title in the subject of your message.

Errata: Although we have taken every care to ensure the accuracy of our content, mistakes do happen. If you have found a mistake in this book, we would be grateful if you would report this to us. Please visit www.packtpub.com/support/errata and fill in the form.

Piracy: If you come across any illegal copies of our works in any form on the internet, we would be grateful if you would provide us with the location address or website name. Please contact us at copyright@packt.com with a link to the material.

If you are interested in becoming an author: If there is a topic that you have expertise in and you are interested in either writing or contributing to a book, please visit authors.packtpub.com.

Share Your Thoughts

Once you've read *Automated Machine Learning on AWS*, we'd love to hear your thoughts! Scan the QR code below to go straight to the Amazon review page for this book and share your feedback.

https://packt.link/r/1801811822

Your review is important to us and the tech community and will help us make sure we're delivering excellent quality content.

Section 1: Fundamentals of the Automated Machine Learning Process and AutoML on AWS

This section will educate you on the complexities of the machine learning process, what AutoML is, and how it can be used to streamline the process.

This section comprises the following chapters:

1
Getting Started with Automated Machine Learning on AWS

If you have ever had the pleasure of successfully driving a production-ready **Machine Learning (ML)** application to completion or you are currently in the process of developing your first ML project, I am sure that you will agree with me when I say, *"This is not an easy task!"*

Why do I say that? Well, if we ignore the intricacies involved in gathering the right training data, analyzing and understanding that data, and then building and training the best possible model, I am sure you will agree that the ML process in itself is a complicated task process, time-consuming, and entirely manual, making it extremely difficult to automate. And it is these factors, plus many more, that contribute to ML tasks being difficult to automate.

The primary goal of this chapter is to emphasize these challenges by reviewing a practical example that sets the stage for why automating the ML process is difficult. This chapter will highlight what governing factors should be considered when performing this automation and how leveraging various **Amazon Web Services** (**AWS**) capabilities can make the task of driving ML projects into production less daunting and fully automated. By the end of this chapter, we will have established a common foundation for overcoming these challenges through automation.

Therefore, in this chapter, we will cover the following topics:

- Overview of the ML process
- Complexities in the ML process
- An example of the end-to-end ML process
- How AWS can make automating ML development and the deployment process easier

Technical requirements

You will need access to the Jupyter Notebook environment to follow along with the example in this chapter. Although sample code has been provided for the various steps of the ML process, a Jupyter Notebook example has been provided in this book's GitHub repository (`https://github.com/PacktPublishing/Automated-Machine-Learning-on-AWS/blob/main/Chapter01/ML%20Process%20Example.ipynb`) for you to work through the entire example at your own pace.

For further instructions on how to set up a Jupyter Notebook environment, you can refer to the installation guide (`https://jupyterlab.readthedocs.io/en/stable/getting_started/installation.html`) to either set up JupyterLab or classic Jupyter Notebook. Alternatively, for local notebook development using a development IDE, such as Visual Studio Code, you can refer to the VS Code documentation (`https://code.visualstudio.com/docs/datascience/jupyter-notebooks`).

Overview of the ML process

Unfortunately, there is no established how-to guide when performing ML. This is because every ML use case is unique and specific to the application that leverages the resultant ML model. Instead, there is a general process pattern that most **data scientists**, **ML engineers**, and **ML practitioners** follow. This process model is called the **Cross-Industry Standard Process for Data Mining** (**CRISP-DM**) and while not everyone follows the specific steps of the process verbatim, most production ML models have probably, in some shape or form, been built by using the guardrails that the CRISP-DM methodology provides.

So, when we refer to the **ML process**, we are invariably referring to the overall methodology of building production-ready ML models using the guardrails from CRSIP-DM.

The following diagram shows an overview of the CRISP-DM guidelines for creating a *typical* process that an ML practitioner might follow:

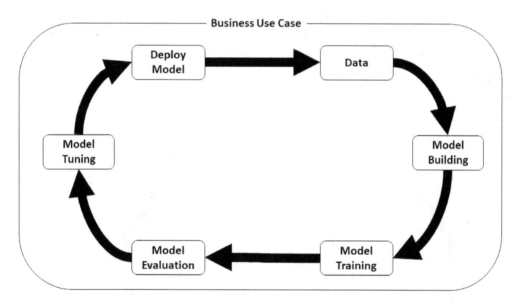

Figure 1.1 – Overview of a typical ML process

In a nutshell, the process starts with the ML practitioner being tasked with providing an ML model that addresses a specific business use case. The ML practitioner then finds, ingests, and analyzes an appropriate dataset that can be effectively leveraged to accomplish the goals of the ML project.

Once the data has been analyzed, the ML practitioner determines the most applicable modeling techniques that extract the most relevant information from the data to address the use case. These techniques include the following:

1. Determining the most applicable ML algorithm
2. Creating new aspects (engineering new features) of the data that can further improve the chosen model's overall effectiveness
3. Separating the data into training and testing sets for model training and evaluation

The ML practitioner then codifies the algorithm's architecture and training/testing/ evaluation routines. These routines are then executed to determine the best possible model parameters – ones that optimize the model to fit both the data and the business use case.

Finally, the best model is deployed into production to serve predictions that match the initial objective of the business use case.

As you can see, the overall process seems relatively straightforward and easy to follow. So, you may be wondering what all the fuss is about. For example, you may be asking yourself, *Where is the complexity in this process?* or *Why do you say that this is so hard to automate?*

While the process may look simplistic, the reality when executing it is vastly different. The following diagram provides a more realistic representation of what an ML practitioner may observe when developing an ML use case:

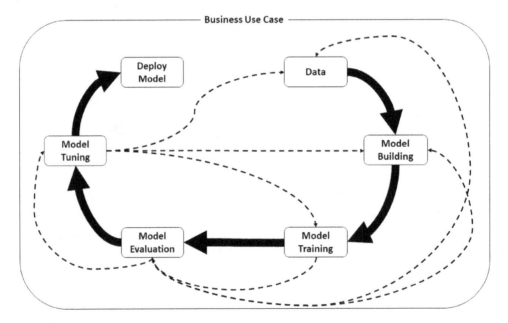

Figure 1.2 – Overview of a realistic ML process

As you can see, the overall process is far more convoluted than the typical representation shown in *Figure 1.1*. There are potentially multiple different paths that can be taken through the process. Each course of action is based on the results captured from the previous step in the process. Additionally, taking a particular course of action may not always yield the desired results, thus forcing the ML practitioner to have to reset or go back and choose a different set of criteria that will hopefully produce a better result.

So, now that we have provided a high-level overview of what the *typical* ML process should entail, let's examine some of the complexities and challenges that make the ML process difficult.

Complexities in the ML process

Each iteration through the process is an experiment to see whether the changes that were made in a previous part of the process will yield a better result or a more optimized ML model. It is this process of **iteration** that makes the ML workflow hard and difficult to automate. The goal of each iteration or experiment is to improve the model's overall predictive capabilities. During each iteration, we fine-tune the parameters, discover new variables, and verify that these changes improve the overall accuracy of the model's prediction. Each experiment also provides further insight into where we are in the overall process and what the next steps might be. In essence, having to potentially go back and tweak a previous step or even go back to the very beginning of the process and start with a different set of data, parameters, or even a different ML model altogether is a manual process. But even unsuccessful experiments have value since they allow us to learn from our mistakes and hopefully steer us toward a successful outcome.

> **Note**
> Tolerating failures and not letting them derail the overall ML process is a key factor in any successful ML strategy.

So, if the overall process is complicated and executing the methodology yields failures, this will hopefully lead to a more successful outcome that will impact the overall ML strategy. It becomes noticeably clear why automating the entire process is challenging but necessary, as it now becomes a crucial part of the overall success criteria of any ML project.

Now that we have a good *idea* of what makes the ML process difficult, let's explore these challenges further by covering a practical example.

An example of the end-to-end ML process

To better illustrate that the overall ML process is hard and that automation is challenging but crucial, we will set the stage with a hands-on example use case.

Introducing ACME Fishing Logistics

ACME Fishing Logistics is a fictitious organization that's concerned with the overfishing of the Sea Snail or Abalone population. Their primary goal is to educate fishermen on how to determine whether an abalone is old enough for breeding. What makes the age determination process challenging is that to verify the abalone's age, it needs to be shucked so that the inside of the shell can be stained and then the number of rings can be counted through a microscope. This involves destroying the abalone to determine whether it is old enough to be kept or returned to the ocean. So, ACME's charter and the goal behind their website is to help fishermen evaluate the various physical characteristics of an abalone so that they can determine its age without killing it.

The case for ML

As you can probably imagine, ACME has not been incredibly successful in its endeavor to prevent abalone overfishing through a simple education process. The CTO has determined that a more proactive strategy must be implemented. Due to this, they have tasked the website manager to make use of ML to make a more accurate prediction of an abalone's age when fishermen enter the physical characteristics of their catch into the new *Age Calculator* module of the website. This is where you come in, as ACME's resident ML practitioner – it is your job to create the ML model that serves abalone age predictions to the new *Age Calculator*.

We can start by using the CRISP-DM guidelines and frame the business use case. The business use case is an all-encompassing step that establishes the overall framework and incorporates the individual steps of the CRISP-DM process.

The purpose of this stage of the process is to establish what the business goals are and to create a project plan that achieves these goals. This stage also includes determining the relevant criteria that define whether, from a business perspective, the project is deemed a success; for example:

- **Business Goal**: The goal of this initiative is to create an *Age Calculator* web application that enables fishermen to determine the age of their abalone catch to determine whether it is below the breeding age threshold. To establish how this business goal can be achieved, several questions arise. For example, how accurate does the age prediction need to be? What evaluation metrics will be used to determine the prediction's accuracy? What is the acceptable accuracy threshold? Is there valid data for the use case? How long will the project take? Having questions like these helps set realistic goals for planning.

- **Project Plan**: A project plan can be formulated by investigating what the answers to some of these questions *might* be. For example, by investigating what data to use and where to find it, we can start to formulate the difficulties in acquiring the data, which impacts how long the project might take. Additionally, understanding about the model's complexity, which also impacts project timelines, as more complicated models require more time to build, evaluate, and tweak.

- **Success Criteria**: As the project plan starts to formulate, we start to get a picture of what success looks like and how to measure it. For example, if we know that creating a complicated model will negatively impact the delivery timeline, we can relax the acceptable prediction accuracy criteria for the model and reduce the time it takes to develop a production-grade model. Additionally, if the business goal is simply to help the fishermen determine the abalone age but we have no way of tracking whether they abide by the recommendation, then our success criteria can be measured – not in terms of the model's accuracy but how often the *Age Calculator* is accessed and used. For instance, if we get 10 application hits a day, then the project can be deemed successful.

While these are only examples of what this stage of the process might look like, it illustrates that careful forethought and planning, along with a very specific set of objectives, must be outlined before any ML processes can start. It also illustrates that this stage of the process cannot be automated, though having a set plan with predefined objectives creates the foundation on which an automation framework could potentially be incorporated.

Getting insights from the data

Now that the overall business case is in place, we can dive into the meat of the actual ML process, starting with the **data** stage. As shown in the following diagram, the data stage is the first individual step within the framework of the business case:

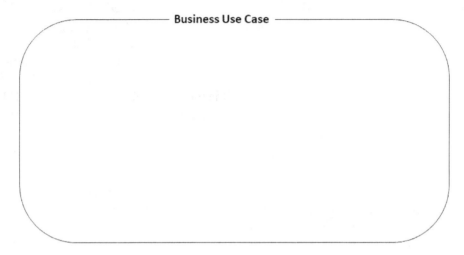

Figure 1.3 – The data stage

It is at this point that we determine what data is available, how to ingest the data, what the data looks like, what characteristics of the data are most relevant to predicting the age, and which features need to be re-engineered to create the most optimal production-ready model.

> **Important Note**
> It is a well-known fact that the data acquisition and exploratory analysis part of the process can account for 70%–80% of the overall effort.

A model worthy of being considered production-ready is only as good as the data it has been trained on. The data needs to be fully analyzed and completely understood to extract the most relevant features for model building and training. We can accomplish this using a technique commonly referred to as **Exploratory Data Analysis** (**EDA**), where we assess the statistical components of the data, potentially visualizing and creating charts to fully grasp feature relevance. Once we have grasped the feature's importance, we might choose to get more important data, remove unimportant data, and potentially engineer new facets of the data, all to have the trained model learn from these optimal features.

Let's walk through an example of what this stage of the process might look like for the *Age Calculator* use case.

Sourcing, ingesting, and understanding the data

For our example, we will be using the *Abalone Dataset*.

> **Note**
>
> The Abalone Dataset is sourced from the University of California, Irvine's ML repository: Dua, D. and Graff, C. (2019). UCI Machine Learning Repository [http://archive.ics.uci.edu/ml]. Irvine, CA: University of California, School of Information and Computer Science.

This dataset contains the various physical characteristics of the abalone that can be used to determine its age. The following steps will walk you through how to access and explore the dataset:

1. We can load the dataset with the following sample Python code, which uses the `pandas` library (`https://pandas.pydata.org`) to ingest the data in a comma-separated value (`csv`) format using the `read_csv()` method. Since the source data doesn't have any column names, we can review the **Attribute Information** section of the dataset website and manually create our `column_names`:

    ```python
    import pandas as pd
    column_names = ["sex", "length", "diameter", "height",
    "whole_weight", "shucked_weight", "viscera_weight",
    "shell_weight", "rings"]
    abalone_data = pd.read_csv("http://archive.ics.uci.edu/
    ml/machine-learning-databases/abalone/abalone.data",
    names=column_names)
    ```

2. Now that the data has been downloaded, we can start analyzing it as a DataFrame. First, we will take a sample of the first five rows of the data to ensure we have successfully downloaded it and verify that it matches the attribute information highlighted on the website. The following sample Python code calls the `head()` method on the `abalone_data` DataFrame:

    ```python
    abalone_data.head()
    ```

The following screenshot shows the output of executing this call:

	sex	length	diameter	height	whole_weight	shucked_weight	viscera_weight	shell_weight	rings
0	M	0.455	0.365	0.095	0.5140	0.2245	0.1010	0.150	15
1	M	0.350	0.265	0.090	0.2255	0.0995	0.0485	0.070	7
2	F	0.530	0.420	0.135	0.6770	0.2565	0.1415	0.210	9
3	M	0.440	0.365	0.125	0.5160	0.2155	0.1140	0.155	10
4	I	0.330	0.255	0.080	0.2050	0.0895	0.0395	0.055	7

Figure 1.4 – The first five rows of the Abalone Dataset

Although we are only viewing the first five rows of the data, it matches the attribute information provided by the repository website. For example, we can see that the **sex** column has nominal values showing if the abalone is male (**M**), female (**F**), or an infant (**I**). We also have the **rings** column, which is used to determine the age of the abalone. The additional columns, such as **weight**, **diameter**, and **height**, detail additional characteristics of the abalone. These characteristics all contribute to determining its age (in years). The age is calculated using the number of rings, plus 1.5.

3.   Next, we can use the following sample code to call the describe() method on the abalone_data DataFrame:

```
abalone_data.describe()
```

The following screenshot shows the summary statistics of the dataset, as well as various statistical details, such as the **percentile**, **mean**, and **standard deviation**:

	length	diameter	height	whole_weight	shucked_weight	viscera_weight	shell_weight	rings
count	4177.000000	4177.000000	4177.000000	4177.000000	4177.000000	4177.000000	4177.000000	4177.000000
mean	0.523992	0.407881	0.139516	0.828742	0.359367	0.180594	0.238831	9.933684
std	0.120093	0.099240	0.041827	0.490389	0.221963	0.109614	0.139203	3.224169
min	0.075000	0.055000	0.000000	0.002000	0.001000	0.000500	0.001500	1.000000
25%	0.450000	0.350000	0.115000	0.441500	0.186000	0.093500	0.130000	8.000000
50%	0.545000	0.425000	0.140000	0.799500	0.336000	0.171000	0.234000	9.000000
75%	0.615000	0.480000	0.165000	1.153000	0.502000	0.253000	0.329000	11.000000
max	0.815000	0.650000	1.130000	2.825500	1.488000	0.760000	1.005000	29.000000

Figure 1.5 – The summary statistics of the Abalone Dataset

> **Note**
>
> At this point, we can gain an understanding of the data by visualizing and plotting any correlations between the key features to further understand how the data is distributed, as well as to determine the most important features in the dataset. We should also determine whether we have missing data and if we have enough data.
>
> Only using summary statistics to understand the data can often be misleading. Although we will not be performing these visualization tasks on this example, you can review why using graphical techniques is so important to understanding data by looking at the **Anscombe's Quartet** example on Kaggle (`https://www.kaggle.com/carlmcbrideellis/anscombe-s-quartet-and-the-importance-of-eda`).

The previous tasks highlight a few important observations we derived from the summary statistics of the dataset. For example, after reviewing the descriptive statistics from the dataset (*Figure 1.5*), we made the following important observations:

- The **count** value for each column is **4177**. We can deduce that we have the same number of observations for each feature and therefore, no missing values. This means that we won't have to somehow infer what these missing values might be or remove the row containing them from the data. Most ML algorithms fail if data is missing.

- If you look at the **75%** value for the **rings** column, there is a significant variance between the **11** rings and that of the **max** amount of rings, which is **29**. This means that the data potentially contains outliers that could add unnecessary *noise* and influence the overall model effectiveness of the trained model.

- While the **sex** column is visible in *Figure 1.4*, the summary statistics displayed in *Figure 1.5* do not include it. This is because of the *type* of data in this column. If you refer to the *Attribute Information* section of the dataset's website (`https://archive.ics.uci.edu/ml/datasets/abalone`), you will see that this **sex** column is comprised of *nominal* data. This type of data is used to provide a label or category for data that doesn't have a quantitative value. Since there is no quantitative value, the summary statistics for this column cannot be displayed. Depending on the type of ML algorithm that's selected to address the business objective, we may need to convert this data into a quantitative format as not all ML algorithms will work with *nominal* data.

The next set of steps will help us apply what we have learned from the dataset to make it more compatible with the model training part of the process:

1. In this step, we focus on converting the **sex** column into quantitative data. The sample code highlights using the `get_dummies()` method on the `abalone_data` DataFrame, which will convert the categories of Male (**M**), Female (**F**), and Infant (**I**) into separate feature columns. Here, the data in these new columns will either reflect one of the categories, represented by a one (**1**) if true or a zero (**0**) if false:

    ```
    abalone_data = pd.get_dummies(abalone_data)
    ```

2. Running the `head()` method again now shows the first five rows of the newly converted data:

    ```
    Abalone_data.head()
    ```

    The following screenshot shows the first five rows of the converted dataset. Here, you can see that the **sex** column has been removed and that, in its place, there are three new columns (one for each new category) with the data now represented as discrete values of **1** or **0**:

	length	diameter	height	whole_weight	shucked_weight	viscera_weight	shell_weight	rings	sex_F	sex_I	sex_M
0	0.455	0.365	0.095	0.5140	0.2245	0.1010	0.150	15	0	0	1
1	0.350	0.265	0.090	0.2255	0.0995	0.0485	0.070	7	0	0	1
2	0.530	0.420	0.135	0.6770	0.2565	0.1415	0.210	9	1	0	0
3	0.440	0.365	0.125	0.5160	0.2155	0.1140	0.155	10	0	0	1
4	0.330	0.255	0.080	0.2050	0.0895	0.0395	0.055	7	0	1	0

Figure 1.6 – The first five rows of the converted Abalone Dataset

3. The next step in preparing the data for model building and training is to separate the **rings** column from the data to establish it as the *target*, or variable, we are trying to predict. The following sample code shows this:

    ```
    y = abalone_data.rings.values
    del abalone_data["rings"]
    ```

4. Now that the target variable has been isolated, we can *normalize* the features. Not all datasets require normalization, however. By looking at *Figure 1.5*, we can see that the summary statistics show that the features have different ranges. These different ranges, especially if the values are large, can influence the overall effectiveness of the model during training. Thus, by normalizing the features, the model can converge to a global minimum much faster. The following code sample shows how the existing features can be normalized by first converting it into a **NumPy** array (https://numpy.org) and then using the normalize() method from the **scikit-learn** or **sklearn** Python library (https://scikit-learn.org/stable/):

```
import numpy as np
from sklearn import preprocessing
X = abalone_data.values.astype(np.float)
X = preprocessing.normalize(X)
```

Based on the initial observations from the dataset, we have applied the necessary transformations to prepare the features for model training. For example, we converted the **sex** column from a *nominal* data type into a quantitative data type since this data will play an important part in determining the age of an abalone.

From this example, you can see that goal of the *Data* step is to focus on exploring and understanding the dataset. We also use this step to apply what we've learned and change the data or **preprocess** it into a representation that suits the downstream model building and training process.

Building the right model

Now that the data has been ingested, analyzed, and processed, we are ready to move onto the next stage of the ML process, where we will look at building the right ML model to suit both the business use case as well as to match it to our newly acquired understanding of the data:

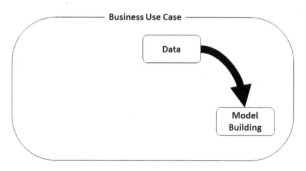

Figure 1.7 – The model building stage

Unfortunately, there is no *one size fits all* algorithm that can be applied to every use case. However, by taking the knowledge we have gleaned from both the business objective and dataset, we can define a list of potential algorithms to use.

For example, we know from our business case that we want to predict the age of the abalone by using the number of rings to get its age. We also know from analyzing and understanding the dataset that we have a target or labeled variable from the **rings** column. This target variable is a discrete, numerical value between 1 and 29, so we can refine our list of possible algorithms to a *supervised learning* algorithm that predicts a *numerical value* among a discrete set of possible values.

The following are just a few of the possible algorithms that could be applied to the example business case:

- Linear regression
- Support vector machines
- Decision trees
- Naïve Bayes
- Neural networks

Once again, there is no one algorithm in this list that perfectly matches the use case and the data. Therefore, the ML process is an experiment to work through multiple possible permutations, get insight from each permutation, and apply what has been learned to further refine the optimal model.

Some of the additional factors that influence which algorithm to start with are based on the ML practitioner's experience, plus how the chosen algorithm addresses the required business goals and success measurements. For example, if a required success criterion is to have the model completed within 2 weeks, then that might eliminate the option to use a more complicated algorithm.

Building a neural network model

Continuing with the *Age Calculator* experiment, we will implement a neural network algorithm, also referred to as **Artificial Neural Network (ANN)**, **Deep Neural Network (DNN)**, or **Multilayer Perceptron (MLP)**.

At a high level, a neural network is an artificial construct modeled on the brain, whereby small, non-linear calculations are made on the data by what is commonly referred to as a **neuron** or **perceptron**. By grouping these neurons into individual layers and then compounding these layers together, we can assemble the building blocks of a mechanism that takes the data as input and finds the dependencies (or correlations) for the output (or target). Through an optimization process, these dependencies are further refined to get the predicted output as close as possible to the actual target value.

> **Note**
>
> The primary reason a neural network model is being used in this example is to introduce a deep learning framework. Deep learning frameworks, such as PyTorch (`https://pytorch.org/`), TensorFlow (`https://www.tensorflow.org/`), and MXNet (`https://mxnet.apache.org/`), can be used to create more complicated neural networks. However, from the perspective of ML process automation, they can also introduce several complexities. So, by making use of a deep learning framework, we can lay the foundation to address some of these complexities later in this book.

The following is a graphical representation of the neural network architecture that we will be building for our example:

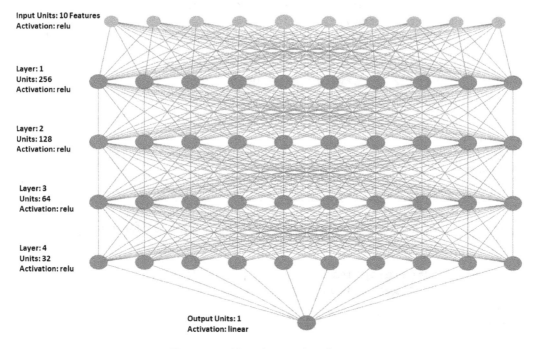

Figure 1.8 – Neural network architecture

The individual components that make up this architecture will be explained in the following steps:

1. To start building the model architecture, we need to load the necessary libraries from the TensorFlow deep learning framework. Along with the `tensorflow` libraries, we will also import the Keras API. The Keras (`https://keras.io/`) library allows us to create higher-level abstractions of the neural network architecture that are easier to understand and work with. For example, from Keras, we also load the `Sequential` and `Dense` classes. These classes allow us to define a model architecture that uses sequential neural network layers and define the type and quantity of neurons in each of these layers:

```
import tensorflow as tf
from tensorflow import keras
from tensorflow.keras.models import Sequential
from tensorflow.keras.layers import Dense
```

2. Next, we can use the `Dense` class to define the list of layers that make up the neural network:

```
network_layers = [
    Dense(256, activation='relu', kernel_
initializer="normal", input_dim=10),
    Dense(128, activation='relu'),
    Dense(64, activation='relu'),
    Dense(32, activation='relu'),
    Dense(1, activation='linear')
]
```

3. Next, we must define the model as being a `Sequential()` model or simply a list of layers:

```
model = Sequential(network_layers)
```

4. Once the model structure has been defined, we must compile it for training using the `compile()` method:

```
model.compile(optimizer="adam", loss="mse",
metrics=["mae", "accuracy"])
```

5. Once the model has been compiled, the `summary()` method can be called to view its architecture:

```
model.summary()
```

The following screenshot shows the results of calling this method. Even though it's showing text output, the network architecture matches the one shown in *Figure 1.8*:

Layer (type)	Output Shape	Param #
dense (Dense)	(None, 256)	2816
dense_1 (Dense)	(None, 128)	32896
dense_2 (Dense)	(None, 64)	8256
dense_3 (Dense)	(None, 32)	2080
dense_4 (Dense)	(None, 1)	33

```
Total params: 46,081
Trainable params: 46,081
Non-trainable params: 0
```

Figure 1.9 – Summary of the compiled neural network architecture

As you can see, the first layer of the model matches **Layer 1** in *Figure 1.8*, where the `Dense()` class is used to express that this layer has 256 neurons, or units, that connect to every neuron in the next layer. **Layer 1** also initializes the parameters (model weights and bias) so that each neuron behaves differently and captures the different patterns we wish to optimize through training. **Layer 1** is also configured to expect input data that has 10 dimensions. These dimensions correspond to the following features of the Abalone Dataset:

- Length
- Diameter
- Height
- Whole Weight
- Shucked Weight

- Viscera Weight
- Shell Weight
- Sex_F
- Sex_I
- Sex_M

Layer 1 is also configured to use the nonlinear **Rectified Linear Unit** (**ReLU**) activation function, which allows the neural network to learn complex relationships from the dataset. We then repeat the process, adding **Layer 2** through **Layer 4**, specifying that each of these layers has 128, 64, 32, and 1 neuron(s) or unit(s), respectively. The final Layer only has a single output – the predicted number of rings. Since the objective of the model is to determine how this output relates to the actual number of rings in the dataset, a linear activation function is used.

Once we have constructed the model architecture, we use the following important parameters to compile the model using the `compile()` method:

- **Loss**: This parameter specifies the type of objective function (also referred to as the cost function) that will be used. At a high level, the objective function calculates how far away or how close the predicted result is to the actual value. It calculates the amount of error between the number of rings that the model predicts, based on the input data, versus what the actual number of rings is. In this example, the **Mean Squared Error** (**MSE**) is used as the objective function, where the average amount of error is measured across all the data points.

- **Optimizer**: The objective during training is to minimize the amount of error between the predicted number of rings and the actual number of rings. The **Adam optimizer** is used to iteratively update the neural network weights that contribute to reducing the loss (or error).

- **Metrics**: The evaluation metrics, **Mean Absolute Error** (**MAE**), and prediction **accuracy** are captured during model training and used to provide insight into how effectively the model is learning from the input data.

> **Note**
>
> If you are unfamiliar with any of these terms, there are a significant amount of references available when you search for them. Additionally, you may find it helpful to take the Deep Learning Specialization course offered by Coursera (`https://www.coursera.org/specializations/deep-learning`). Further details on these parameters can be found in the Keras API documentation (`https://keras.io/api/models/model_training_apis/#compile-method`).

Now that we have built the architecture for the neural network algorithm, we need to see how it *fits* on top of the preprocessed dataset. This task is commonly referred to as *training* the model.

Training the model

The next step of the ML process, as illustrated in the following diagram, is to train the dataset on the preprocessed abalone data:

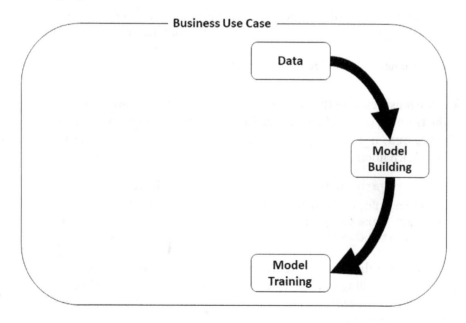

Figure 1.10 – The model training stage

Training the compiled model is relatively straightforward. The following steps outline how to kick off the model training part of the process:

1. This first step is not necessary to train the model, but sometimes, the output from the training process can be unwieldy and difficult to interpret. Therefore, a custom class called `cleanPrint()` can be created to ensure that the training output is neat. This class uses the Keras `Callback()` method to print a dash (`"-"`) as the training output:

    ```
    class cleanPrint(keras.callbacks.Callback):
        def on_epoch_end(self, epoch, logs):
            if epoch+1 % 100 == 0:
                print("!")
            else:
                print("-", end="")
    ```

    > **Note**
    >
    > It is a good practice to display the model's performance at each epoch as this provides insight into the improvements after each epoch. However, since we are training for 2000 epochs, we are using the `cleanPrint()` class to make the output neater. We will remove this callback later.

2. Next, we must separate the preprocessed abalone data into two main groups – one for the **training** data and one for **testing** data. The splitting process is performed by using the `train_test_split()` method from the `model_selection()` class of the sklearn library:

    ```
    from sklearn.model_selection import train_test_split
    training_features, testing_features, training_labels,
    testing_labels = train_test_split(X, y, test_size=0.2,
    random_state=42)
    ```

3. The final part of the training process is to launch the model training process. This is done by calling the `fit()` method on the compiled model and supplying the `training_features` and `training_labels` datasets, as shown in the following example code:

    ```
    training_results = model.fit(training_features, training_
    labels, validation_data=(testing_features, testing_
    labels), batch_size=32, epochs=2000, shuffle=True,
    verbose=0, callbacks=[cleanPrint()])
    ```

Now that the model training process has started, we can review a few key aspects of our code. First, splitting the data into training and testing datasets is typically performed as part of the data preprocessing step. However, we are performing this task during the model training step to provide additional context to the loss and optimization functions. For example, creating these two separate datasets is an important part of evaluating how well the model is being trained. The model is trained using the training dataset and then its effectiveness is evaluated against the testing dataset. This evaluation procedure guides the model (using the **loss function** and the **optimization function**) to reduce the amount of error between the predicted number of rings and the actual number of rings. In essence, this makes the model better or optimizes the model. To create a good split of training and testing data, we must provide four additional variables, as follows:

- `training_features`: The 10 columns of the Abalone Dataset that correspond to the abalone attributes, comprising 80% of these observations.

- `testing_features`: The same 10 columns of the Abalone Dataset, comprising the other 20% of the observations.

- `training_labels`: The number of rings (target label) for each observation in the `training_features` dataset.

- `testing_labels`: The number of rings (target label) for each observation in the `testing_features` dataset.

> **Tip**
> Further details about each of these parameters, as well as more parameters that you can use to tweak the training process, can be found in the Keras API documentation (`https://keras.io/api/models/model_training_apis/#fit-method`).

Secondly, once the data has been successfully split, we can use the `fit()` method and add the following parameters to further govern the training process:

- `validation_data`: The `testing_features` and `testing_labels` datasets, which the model uses to evaluate how well the trained neural network weights reduce the amount of error between the predicted number of rings and the actual number of rings in the testing data.

- `batch_size`: This parameter defines the number of samples from the training data that are propagated through the neural network. This parameter can be used to influence the overall speed of the training process. The higher `batch_size` is, the higher the number of samples that are used from the training data, which means the higher the number of samples that are combined to estimate the loss before updating the neural network's weights.

- `epochs`: This parameter defines how many times the training process will iterate through the training data. The higher `epochs` is, the more iterations must be made through the training data to optimize the neural network's weights.

- `shuffle`: This parameter specifies whether to shuffle the data before starting a training iteration. Shuffling the data each time the model iterates through the data forces the model to generalize better and prevent it from learning ordered patterns in the training data.

- `verbose` and `callbacks`: These parameters are related to displaying the training progress and output for each `epoch`. Setting the output to zero and using the `cleanPrint()` class will simply display a dash (-) as the output for each `epoch`.

The training process should take 12 minutes to complete, providing us with a trained `model` object. In the next section, we will use the trained model to evaluate how well it makes predictions on new data.

Evaluating the trained model

Once the model has been trained, we can move on to the next stage of the ML process: the model evaluation stage. It is at this stage that the trained model is evaluated against the objectives and success criterion that have been established within the business use case, with the goal being to determine if the trained model is ready for production or not:

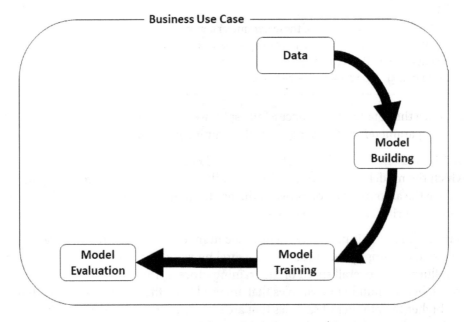

Figure 1.11 – The model evaluation step

When evaluating a trained model, most ML practitioners simply score the quality of the model predictions using an evaluation metric that is suited to the type of model. Other ML practitioners go one step further to visualize and further understand the predictions. The following steps will walk you through using the latter of these two approaches:

1. Using the following sample code, we can load the necessary Python libraries. The first library is `matplotlib`. The `pyplot()` class is a collection of different functions that allow for interactive and programmatic plot generation. The second library, `mean_squarred_error()`, comes from the `sklearn` package and provides the ML practitioner with an easy way to evaluate the quality of the model using the **Root Mean Squared Error** (**RMSE**) metric. Since the neural network model is a supervised learning-based regression model, RMSE is a popular method that's used to measure the error rate of the model predictions:

```python
import matplotlib.pyplot as plt
from sklearn.metrics import mean_squared_error
```

2. The imported libraries are then used to visualize the predictions to provide a better understanding of the model's quality. The following code generates a plot that incorporates the information that's required to quantify the prediction's quality:

```python
fig, ax = plt.subplots(figsize=(15, 10))
ax.plot(testing_labels, model.predict(testing_features),
"ob")
ax.plot([0, 25], [0, 25], "-r")
ax.text(8, 1, f"RMSE = {mean_squared_error(testing_
labels, model.predict(testing_features),
squared=False)}", color="r", fontweight=1000)
plt.grid()
plt.title("Abalone Model Evaluation", fontweight="bold",
fontsize=12)
plt.xlabel("Actual 'Rings'", fontweight="bold",
fontsize=12)
plt.ylabel("Predicted 'Rings'", fontweight="bold",
fontsize=12)
plt.legend(["Predictions", "Regression Line"], loc="upper
left", prop={"weight": "bold"})
plt.show()
```

Executing this code will create two sub-plots. The first sub-plot is a scatterplot displaying the model predictions from the test dataset, as well as the ground truth labels. The second sub-plot superimposes a regression line over these predictions to highlight the linear relationship between the predicted number of rings versus the actual number of rings. The rest of the code labels the various properties of the plot and displays the RMSE score of the predictions. The following is an example of this plot:

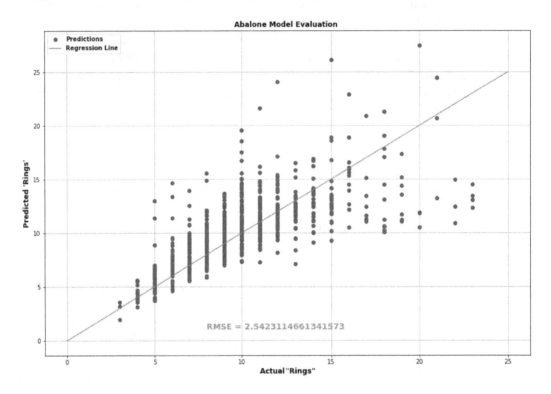

Figure 1.12 – An example Abalone Model Evaluation scatterplot

Three things should immediately stand out here:

- The RMSE evaluation metric scores the trained model at 2.54.

- The regression line depicting the correlation between the actual number of rings and the predicted number of rings does not pass through the majority of the predictions.

- There are a significant number of predictions that are far away from the regression line on both the positive and negative scales. This shows a high error rate between the number of rings that are predicted versus the actual number of rings for a data point.

These observations and others should be compared to the objectives and success criteria that are outlined in the business use case. Both the ML practitioner and business owner can then judge whether the trained model is ready for production.

For example, if the primary objective of the *Age Calculator* application is to use the model predictions as a rough guide for the fishermen to get a simple idea of the abalone age, then the model does this and can therefore be considered ready for production. If, on the other hand, the primary goal of the *Age Calculator* application is to provide an accurate age prediction, then the example model probably cannot be considered production-ready.

So, if we determine that the model is not ready for production, *what are the subsequent steps of the ML process?* The next section will review some options.

Exploring possible next steps

Since the model has been deemed unfit for production, several approaches can be taken after the model evaluation stage. The following diagram highlights three possible options that can be considered as possible next steps:

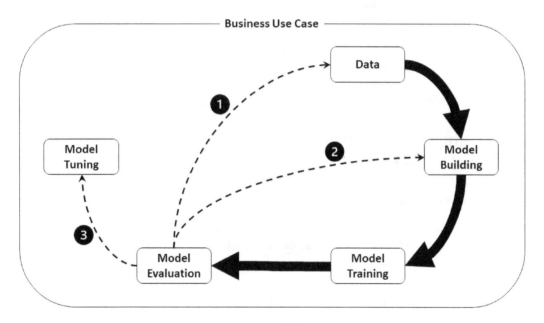

Figure 1.13 – Next step options

Let's explore these three possible next steps in more depth to determine which option best suits the objectives of the *Age Calculator* use case.

Option 1 – get mode data

The first option requires the ML practitioner to go back to the beginning of the process and acquire more data. Since the UCI abalone repository is the only publicly available dataset, this task might involve physically gathering more observations by manually fishing for abalone or conducting a survey with fishermen on their catch. Either way, this takes time!

However, simply adding more observations to the dataset does not necessarily translate to a better-quality model. So, getting more data could also mean getting better-quality features. This means that the ML practitioner would need to reevaluate the existing data, dive further into the analysis to better understand which of the features are of the most importance, and then re-engineer those features or create new features from them. *This too is time-consuming*!

Option 2 – choose another model

The second option to consider involves building an entirely new model using a completely different algorithm that still matches the use case. For example, the ML practitioner might investigate using another supervised learning, regression-based algorithm.

Different algorithms might also require the data to be restructured so that it's more suited to the algorithm's required type of input. For example, choosing a **Gradient Boosting Regression** algorithm, such as **XGBoost**, requires the target label to be the first column in the dataset. *Choosing another algorithm and reengineering the data requires additional time!*

Option 3 – tuning the existing model

Recall that when the existing neural network model was built, there were a few tunable parameters that were configured during its compilation. For example, the model was compiled using particular optimizer and loss functions.

Additionally, when the existing neural network model was trained, other tunable parameters were supplied, such as the number of **epochs** and the **batch size**.

> **Note**
>
> There is no best practice for choosing the right option. Remember that each iteration through the process is an experiment whereby the goal is to glean more information from the experiment to determine the next course of action or next option.

While *Option 3* may seem straightforward, in the next section, you will see that this option also involves multiple potential iterations and is therefore also *time-consuming*.

Tuning our model

As we've already highlighted, multiple parameters or hyperparameters can be tuned to better tune or optimize an existing model. Hence, this stage of the process is also referred to as **hyperparameter optimization**. The following diagram shows what the hyperparameter optimization process entails:

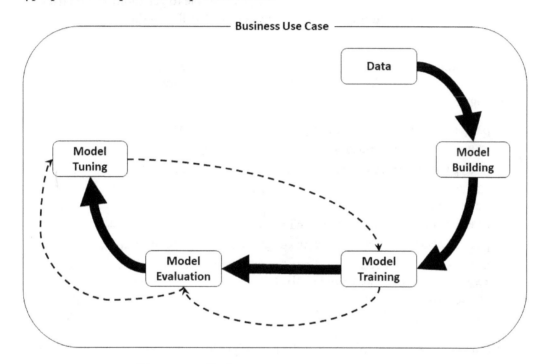

Figure 1.14 – The hyperparameter optimization process

After evaluating the model to determine which hyperparameters can be tweaked, the model is trained using these parameters. The trained model is, once again, compared to the business objectives and success criterion to determine if it is ready for production. This process is then repeated, constantly tweaking, training, and evaluating until a production-ready model is produced.

Determining the best hyperparameters to tune

Once again, there is no exact approach to getting the optimal hyperparameters. Each iteration through the process helps narrow down which combination of hyperparameters contributes to a more optimized model.

However, a good place to start the process is to dive deeper into what is happening during model training and derive further insights into how the model is learning from the data.

You will recall that, when executing the `fit()` method to train the model and by binding the results to the `training_results` parameter, we are able to get additional metrics that were needed for model tuning. The following steps will walk you through an example of how to extract and visualize these metrics:

1. By using the `history()` method on the `training_results` parameter, we can use the following sample code to plot the prediction error for both the training and testing processes.

    ```
    plt.rcParams["figure.figsize"] = (15, 10)
    plt.plot(training_results.history["loss"])
    plt.plot(training_results.history["val_loss"])
    plt.title("Training vs. Testing Loss", fontweight="bold",
    fontsize=14)
    plt.ylabel("Loss", fontweight="bold", fontsize=14)
    plt.xlabel("Epochs", fontweight="bold", fontsize=14)
    plt.legend(["Training Loss", "Testing Loss"], loc="upper
    right", prop={"weight": "bold"})
    plt.grid()
    plt.show()
    ```

    The following is an example of what the plot might look like after executing the preceding code:

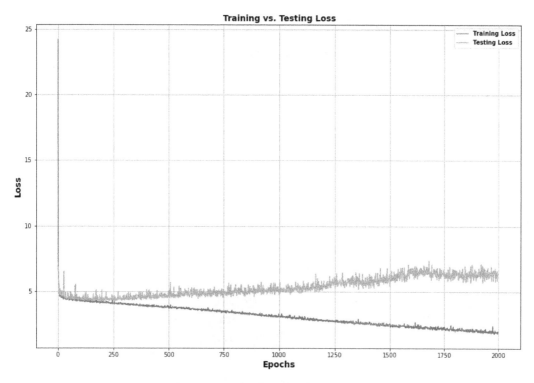

Figure 1.15 – Training vs. Testing Loss

2. Similarly, by replacing the loss and val_loss parameters in the sample code with mae and val_mae, respectively, we can see a consistent trend:

```
plt.rcParams["figure.figsize"] = (15, 10)
plt.plot(training_results.history["mae"])
plt.plot(training_results.history["val_mae"])
plt.title("Training vs. Testing Mean Absolute Error",
fontweight="bold", fontsize=14)
plt.ylabel("mae", fontweight="bold", fontsize=14)
plt.xlabel("Epochs", fontweight="bold", fontsize=14)
plt.legend(["Training MAE", "Testing MAE"], loc="upper
right", prop={"weight": "bold"})
plt.grid()
plt.show()
```

After executing the preceding code, we will get the following output:

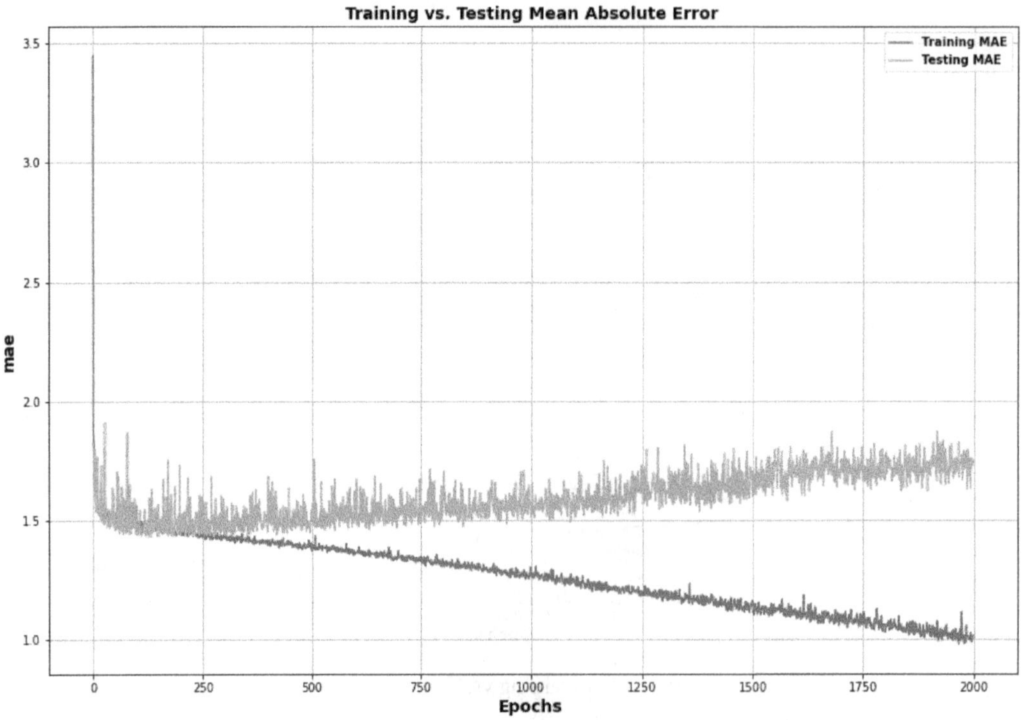

Figure 1.16 – Training vs. Testing Mean Absolute Error

Both *Figure 1.16* and *Figure 1.15* clearly show a few especially important trends:

- There is a clear divergence between what the model is learning from the training data and its predictions on the testing data. This indicates that the model is not learning anything new as it trains and is essentially overfitting the data. The model relates to the training data and is unable to relate to new, unseen data in the testing dataset.

- This divergence seems to happen around 250 epochs/training iterations. Since the training process was set to 2,000 epochs, this indicates that the model is being over-trained, which could be the reason it is overfitting the training data.

- Both the testing MAE and the testing loss have an erratic gradient. This means that as the model parameters are being updated through the training process, the magnitude of the updates is too large, resulting in an unstable neural network, and therefore unstable predictions on the testing data. So, the fluctuations depicted by the plot essentially highlight an **exploding gradient** problem, indicating that the model is overfitting the data.

Based on these observations, several hyperparameters can be tuned. For example, an obvious parameter to change is the number of epochs or training iterations to prevent overfitting. Similarly, we could change the optimization function from **Adam** to **Stochastic Gradient Descent (SGD)**. SGD allows a specific learning rate to be set as one of its parameters, as opposed to the adaptive learning rate used by the Adam optimizer. By specifying a small learning rate parameter, we are essentially rescaling the model updates to ensure that they are small and controlled.

Another solution might be to use a regularization technique, such as *L1* or *L2* regularization, to penalize some of the neurons on the model, thus creating a simpler neural network. Likewise, simplifying the neural network architecture by reducing the number of layers and neurons within each layer would have the same effect as regularization.

Lastly, reducing the number of samples or batch size can control the stability of the gradient during training.

Now that we have a fair idea of which hyperparameters to tweak, the next section will show you how to further optimize the model.

Tuning, training, and reevaluating the existing model

We can start model tuning by walking through the following steps:

1. The first change we must make is to the neural network architecture itself. The following example code depicts the new structure, where only two network layers are used instead of four. Each layer only has 64 neurons:

```
network_layers = [
    Dense(64, activation='relu', kernel_
initializer="normal", input_dim=10),
    Dense(64, activation='relu'),
    Dense(1, activation='linear')
]
```

2. Once again, the model is recompiled using the same parameters as those from the previous example:

```
model = Sequential(network_layers)
model.compile(optimizer="adam", loss="mse",
metrics=["mae", "accuracy"])
model.summary()
```

The following screenshot shows the text summary of the tuned neural network architecture:

Layer (type)	Output Shape	Param #
dense_5 (Dense)	(None, 64)	704
dense_6 (Dense)	(None, 64)	4160
dense_7 (Dense)	(None, 1)	65

Total params: 4,929
Trainable params: 4,929
Non-trainable params: 0

Figure 1.17 – Summary of the tuned neural network architecture

The following diagram shows a visual representation of the turned neural network architecture:

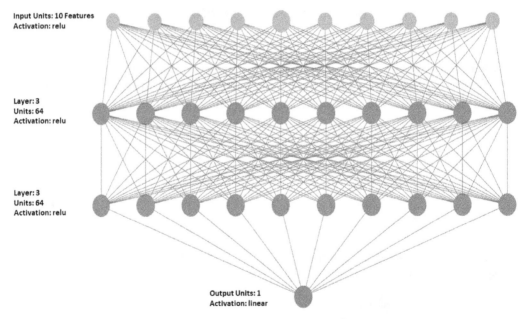

Figure 1.18 – Tuned neural network architecture

3. Lastly, the `fit()` method is called on the new model. However, this time, the number of epochs has been reduced to `200` and `batch_size` has also been reduced to `8`:

```
training_results = model.fit(training_features,
training_labels, validation_data=(testing_features,
testing_labels), batch_size=8, epochs=200, shuffle=True,
verbose=1)
```

> **Note**
>
> In the previous code example, the `cleanPrint()` callback has been removed to show the evaluation metrics on both the training and validation data at 200 epochs.

4. Once the new model training has been completed, the previously used evaluation code can be re-executed to display the evaluation scatterplot. The following is an example of this scatterplot:

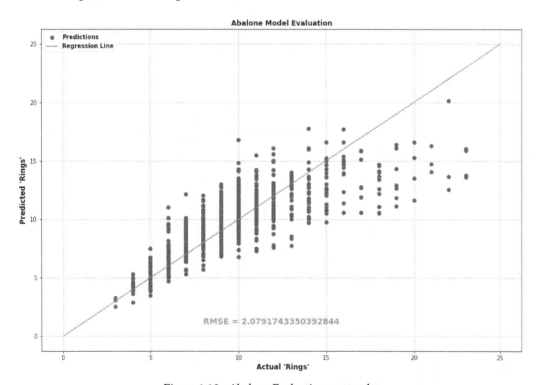

Figure 1.19 – Abalone Evaluation scatterplot

The new model does not capture all the predictions as there are still several outliers on the positive and negative scales. However, there is a drastic improvement to the overall fit on most data points. This is further quantified by the RMSE score dropping from **2.54** to **2.08**.

Once again, these observations should be compared to the objectives and the success criteria that are outlined in the business use case to gauge whether the model is ready for production.

As the following diagram illustrates, if a production-ready model cannot be found, then the options to further tune the model, get and engineer more data, or build a completely different model are still available:

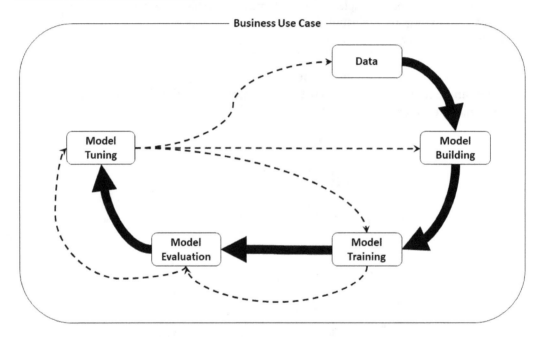

Figure 1.20 – Additional process options

Should the model be deemed as production-ready, the ML practitioner can move onto the final stage of the ML process, As shown in the following diagram this is the model deployment stage:

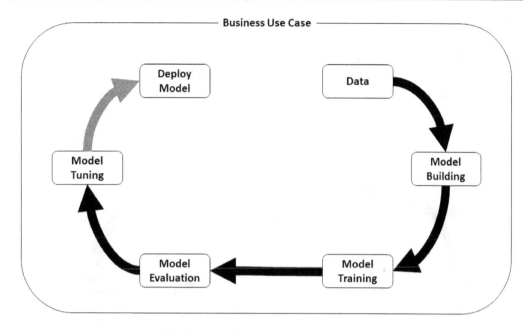

Figure 1.21 – The model deployment stage

In the next section, we will review the processes involved in deploying the model into production.

Deploying the optimized model into production

Model deployment is somewhat of a gray area in that some ML practitioners do not apply this stage to their ML process. For example, some ML practitioners may feel that the scope of their task is to simply provide a production-ready ML model that addresses the business use case. Once this model has been trained, they simply hand it over to the application development teams or application owners for them to test and integrate the model into the application.

Alternatively, some ML practitioners will work with the application teams to deploy the model into a **test** or **Quality Assurance (QA)** environment to ensure that the trained model successfully integrates with the application.

Whatever the scope of the ML practitioner role, model deployment is part of the CRISP-DM methodology and should always be factored into the overall ML process, especially if the ML process is to be automated.

While the CRISP-DM methodology ends with the model deployment stage, as shown in the preceding diagram, the process is, in fact, a continuous process. Once the model has been deployed into a production application, it needs to be constantly monitored to ensure that it does not drift from its intended purpose, to consistently provide accurate predictions on unseen data or new data. Should this situation arise, the ML practitioner will be called upon to start the ML process again to reoptimize the model and make it generalize to this new data. The following diagram shows what the ML process looks like in reality:

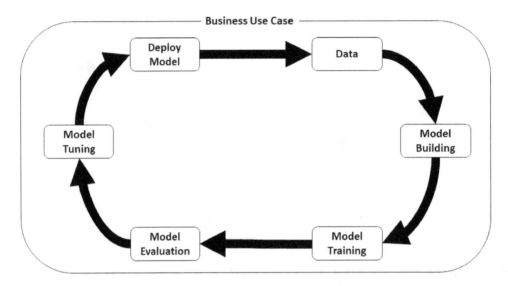

Figure 1.22 – Closing the loop

So, once again, *why is the ML process hard?*

Using this simple example use case, you can hopefully see that not only are there inherent complexities to the process of exploring the data, as well as building, training, evaluating, tuning, deploying, and monitoring the model – the entire process is also complex, manual, iterative, and continuous.

How can we streamline the process to ensure that the outcome is always an optimized model that matches the business use case? This is where **AutoML** comes into play.

Streamlining the ML process with AutoML

AutoML is a broad term that has different a meaning depending on who you ask. When referring to AutoML, some ML practitioners may point to a dedicated software application, a set of tools/libraries, or even a dedicated cloud service. In a nutshell, AutoML is a methodology that allows you to create a repeatable, reliable, streamlined, and, of course, automated ML process.

The process is **repeatable** in that it follows the same pattern every time it is executed. The process is **reliable** in that it guarantees that an optimized model that matches the use case is always produced. The process is **streamlined** and any unnecessary steps are removed, making it as efficient as possible. Finally, and most importantly, the process can be started and executed automatically and triggered by an event, such as retraining the model after model concept drift has been detected.

AWS provides multiple capabilities that can be used to build a streamlined AutoML process. In the next section, I will highlight some of the dedicated cloud services, as well as other services, that can be leveraged to make the ML process easier and automated.

How AWS makes automating the ML development and deployment process easier

The focus of the remaining chapters in this book will be to practically showcase, using hands-on examples, how the ML process can be automated on AWS. By expanding on the *Age Calculator* example, you will see how various AWS capabilities and services can be used to do this. For example, the next two chapters of this book will focus on how to use some of the native capabilities of the AWS AI/ML stack, such as the following:

- Using SageMaker Autopilot to automatically create, manage, and deploy an optimized abalone prediction model using both codeless as well as coded methods.

- Using the AutoGluon libraries to determine the best deep learning algorithm to use for the abalone model, as well as an example for more complicated ML use cases, such as computer vision.

Parts two, three, and four of this book will focus on leveraging other AWS services that are not necessarily part of the AI/ML stack, such as the following:

- AWS CodeCommit and CodePipeline, which will deliver the abalone use case using a **Continuous Integration and Continuous Delivery (CI/CD)** pipeline.

- AWS Step Functions and the Data Science Python SDK, to create a codified pipeline to produce the abalone model.

- Amazon **Managed Workflows for Apache Airflow (MWAA)**, to automate and manage the ML process.

Finally, part five of this book will expand on some of the central topics that were covered in parts two and three to provide you with a hands-on example of how a cross-functional, agile team can implement the end-to-end *Abalone Calculator* example as part of a **Machine Learning Software Development Life Cycle (MLSDLC)**.

Summary

As I stated from the outset, the primary goal of this chapter was to emphasize the many challenges an ML practitioner may face when building an ML solution for a business use case. In this chapter, I introduced you to an example ML use case – the *Abalone Calculator* – and I used it to show you just how hard the ML process is in reality.

By walking through each step of the process, I explained the complexities involved therein, as well as the challenges you could potentially encounter. I also highlighted why the ML process is complicated, manual, iterative, and continuous, which set the stage for an automated process that is repeatable, streamlined, and reliable using AutoML.

In the next chapter, we will explore how to start implementing an AutoML methodology by introducing you to a native AWS service called SageMaker Autopilot.

2
Automating Machine Learning Model Development Using SageMaker Autopilot

AWS offers a number of approaches for automating ML model development. In this chapter, I will present one such method, **SageMaker Autopilot**. Autopilot is a framework that automatically executes the key steps of a typical ML process. This allows both the novice, as well as the experienced ML practitioner to delegate the manual tasks of data exploration, algorithm selection, model training, and model optimization to a dedicated AWS service, basically, automating the end-to-end ML process.

Before we can start diving and getting hands-on exposure to the native capabilities that AWS offers for ML process automation, it is important to first understand the landscape of where they fit, what these capabilities are, and how we will use them.

In this chapter, we will introduce you to some of the AWS capabilities that focus on ML solutions, as well as ML automation. By the end of the chapter, you will have a hands-on overview of how to automate the *ACME Fishing Logistics* use case using AWS services to implement an AutoML methodology. We will be covering the following topics:

- Introducing the AWS AI and ML landscape
- Overview of SageMaker Autopilot
- Overcoming automation challenges with SageMaker Autopilot
- Using the SageMaker SDK to automate the ML experiment

Technical requirements

You should have the following prerequisites before getting started with this chapter:

- Familiarity with AWS and its basic usage.
- A web browser (for the best experience, it is recommended that you use the Chrome or Firefox browser).
- An AWS account (if you are unfamiliar with how to get started with an AWS account, you can go to this link: `https://aws.amazon.com/getting-started/`).
- Familiarity with the AWS Free Tier (the Free tier will allow you to access some of the AWS services for free, depending on resource limits; you can familiarize yourself with these limits at this link: `https://aws.amazon.com/free/`).
- Example Jupyter notebooks for this chapter are provided in the companion GitHub repository (`https://github.com/PacktPublishing/Automated-Machine-Learning-on-AWS/blob/main/Chapter02/Autopilot%20Example.ipynb`).

Introducing the AWS AI and ML landscape

AWS provides its customers with an extensive assortment of **Artificial Intelligence** (**AI**) and ML capabilities. To further help its customers to better understand these capabilities, AWS has grouped and organized them together into what is typically referred to as the **AI/ML Stack**. The primary goal behind the AI/ML stack is to provide the necessary resources that a developer or ML practitioner might use, depending on their level of expertise. Basically, it puts AI and ML capabilities into the hands of every developer, no matter whether they are considered a novice or an expert. *Figure 2.1* shows the layers that comprise the AWS AI/ML stack.

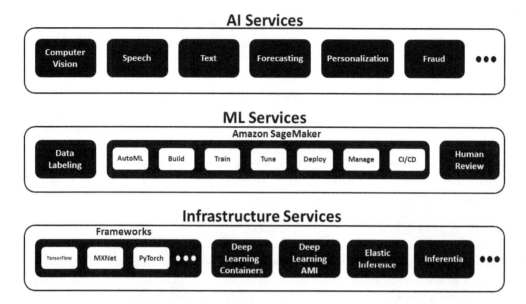

Figure 2.1 – Layers of the AWS AI/ML stack

As you can see from *Figure 2.1*, the AI/ML stack delivers on the goal of putting AI and ML capabilities into the hands of every developer, by grouping the AWS capabilities into three specific layers, where each layer comprises the typical AWS AI/ML resources that meet both the use case requirements and the practitioner's level of comfort and expertise.

For example, should an expert ML practitioner desire to build their own model training and hosting architecture using Kubernetes, then the bottom layer of the AI/ML stack will provide them with all the AWS resources they will need to build this infrastructure. Resources such as dedicated **Elastic Compute Cloud** (**EC2**) instances with all the ML libraries pre-packaged as **Amazon Machine Images** (**AMIs**) for both CPU-based and GPU-based model training and hosting. Hence the bottom-most layer of the AI/ML stack requires a high degree of expertise, as the ML practitioner has the most flexibility, but also the most difficult task of creating their own ML infrastructure to address the ML use case.

Alternatively, should a novice ML practitioner need to deliver an ML model that addresses a specific use case, such as **Object Detection** in images and video, and they don't have the necessary expertise, then the top layer of the AI/ML stack will provide them with the AWS resources to accomplish this. For instance, one of the dedicated AI services within the top layer, **Amazon Rekognition**, provides a pre-built capability to identify objects in both images and video. This means that the ML practitioner can simply integrate the Rekognition service into their production application without having to build, train, optimize, or even host their own ML model. So, by using these applied AI services in the top layer of the stack, details about the model, the training data used, or which hyperparameters were tuned are abstracted away from the user. Consequently, these applied AI services are easier to use and provide a faster mean time to delivery for the business use case.

As we go down the stack, we see that the ML practitioner is responsible for configuring specific details about the model, the tuned hyperparameters, and the training datasets. So, to help their customers with these tasks, AWS provides a dedicated service at the middle layer of the AI/ML stack, called Amazon SageMaker. SageMaker fits comfortably into the middle in that it caters to experienced ML practitioners by providing them with the flexibility and functionality to handle complex ML use cases without having to build and maintain any infrastructure. From the perspective of the novice ML practitioner, SageMaker allows them to use its built-in capabilities to easily build, train, and deploy simple and advanced ML use cases.

Even though SageMaker is a single AWS service, it has several capabilities or modules that take care of all the heavy lifting for each step of the ML process. Both novice and experienced ML practitioners can leverage the integrated ML development environment (SageMaker Studio) or the Python SDK (SageMaker SDK) to explore and wrangle large quantities of data and then build, train, tune, deploy, and monitor their ML models at scale. *Figure 2.2* shows how some of these SageMaker modules map to and scale each step of the ML process:

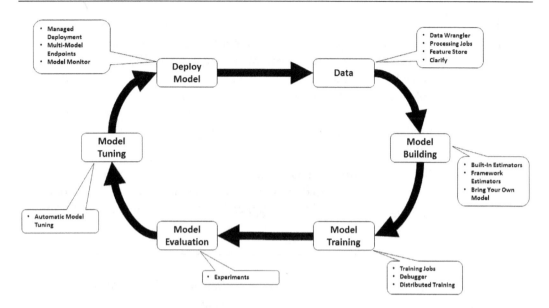

Figure 2.2 – Overview of SageMaker's capabilities

We will not be diving deeper into the capabilities highlighted in *Figure 2.2*, as we will be leveraging some of the features in later chapters and, as such, we will be explaining how they work then. For now, let's dive deeper into the native SageMaker module responsible for automating the ML process, called SageMaker Autopilot.

Overview of SageMaker Autopilot

SageMaker Autopilot is the AWS service that provides AutoML functionality to its customers. Autopilot addresses the various requirements for AutoML by piecing together the following SageMaker modules into an automated framework:

- **SageMaker Processing**: Processing jobs take care of the heavy lifting and scaling requirements of organizing, validating, and feature engineering the data, all using a simplified and managed experience.

- **SageMaker Built-in Algorithms**: SageMaker helps ML practitioners to get started with model-building tasks by providing several pre-built algorithms that cater to multiple use case types.

- **SageMaker Training**: Training jobs take care of the heavy lifting and scaling tasks associated with provisioning the required compute resources to train the model.

- **Automatic Model Tuning**: Model tuning or hyperparameter tuning scales the model tuning task by allowing the ML practitioner to execute multiple training jobs, each with a subset of the required parameters, in parallel. This removes the iterative task of having to sequentially tune, evaluate, and re-train the model. By default, SageMaker model tuning uses **Bayesian Search** (**Random Search** can also be configured) to essentially create a probabilistic model of the performance of previously used hyperparameters to select future hyperparameters that better optimize the model.

- **SageMaker Managed Deployment**: Once an optimized model has been trained, SageMaker Hosting can be used to deploy either a single model or multiple models as a fully functioning API for production applications to consume in an elastic and scalable fashion.

> **Tip**
>
> For more information on these SageMaker modules, you can refer to the AWS documentation (`https://docs.aws.amazon.com/sagemaker/latest/dg/whatis.html`).

Autopilot links these capabilities together, to create an automated workflow. The only piece that the ML practitioner must supply is the raw data. Autopilot, therefore, makes it easy for even the novice ML practitioner to automatically create a production-ready model, just by simply supplying the data.

Let's get started with Autopilot so you can see for yourself just how easy this process really is.

Overcoming automation challenges with SageMaker Autopilot

In *Chapter 1, Getting Started with Automated Machine Learning on AWS*, we practically highlighted the challenges that ML practitioners face when creating production-ready ML models. By way of a recap, these challenges are grouped into two main categories:

- The challenges imposed by building the best ML model, such as sourcing and understanding the data and then building the best model for the use case

- The challenges imposed by the ML process itself, the fact that it is complicated, manual, iterative, and continuous

So, in order to better understand just how Autopilot overcomes these challenges, we must understand the anatomy of the Autopilot workflow and how it compares to the example ML process we discussed in *Chapter 1, Getting Started with Automated Machine Learning on AWS*.

Before we begin to use Autopilot, we need to understand that there are multiple ways to interface with the service. For example, we can use the AWS **Command-Line Interface (CLI)**, call the service **Application Programing Interface (API)** programmatically using the **Software Development Kits (SDKs)**, or simply use the **SageMaker Python SDK**. However, Autopilot offers an additional, easy-to-use interface that is incorporated into SageMaker Studio. We will use the **SageMaker Studio IDE** for this example.

The following section will walk you through applying an AutoML methodology to the *Abalone Calculator* use case, with SageMaker Autopilot.

Getting started with SageMaker Studio

Depending on your personal or organizational usage requirements, SageMaker Studio offers multiple ways to get started. For example, should an ML practitioner be working as part of a team, **AWS Single Sign-On (SSO)** or **AWS Identity & Access** users can be configured for the team. However, for this example use case, we will onboard to Studio using the QuickStart procedure as it is the most convenient for individual user access. The following steps will walk you through setting up the Studio interface:

1. Log into your AWS account and select an AWS region where SageMaker is supported.

    > **Note**
    >
    > If you are unsure which AWS regions support SageMaker, refer to the following link: SageMaker Supported Regions and Quotas (`https://docs.aws.amazon.com/sagemaker/latest/dg/regions-quotas.html`).

2. Navigate to the SageMaker service console by entering `SageMaker` in the search bar, or by clicking on **Amazon SageMaker** from the **Services** dropdown.

3. Using the left-hand navigation panel, click **Studio**, under the **SageMaker Domain** option.

4. Since this is the first time a SageMaker Studio domain is being configured, the **Setup SageMaker Domain** console will offer two setup options, namely **Quick setup** and **Standard setup**. *Figure 2.3* shows what the screen should look like:

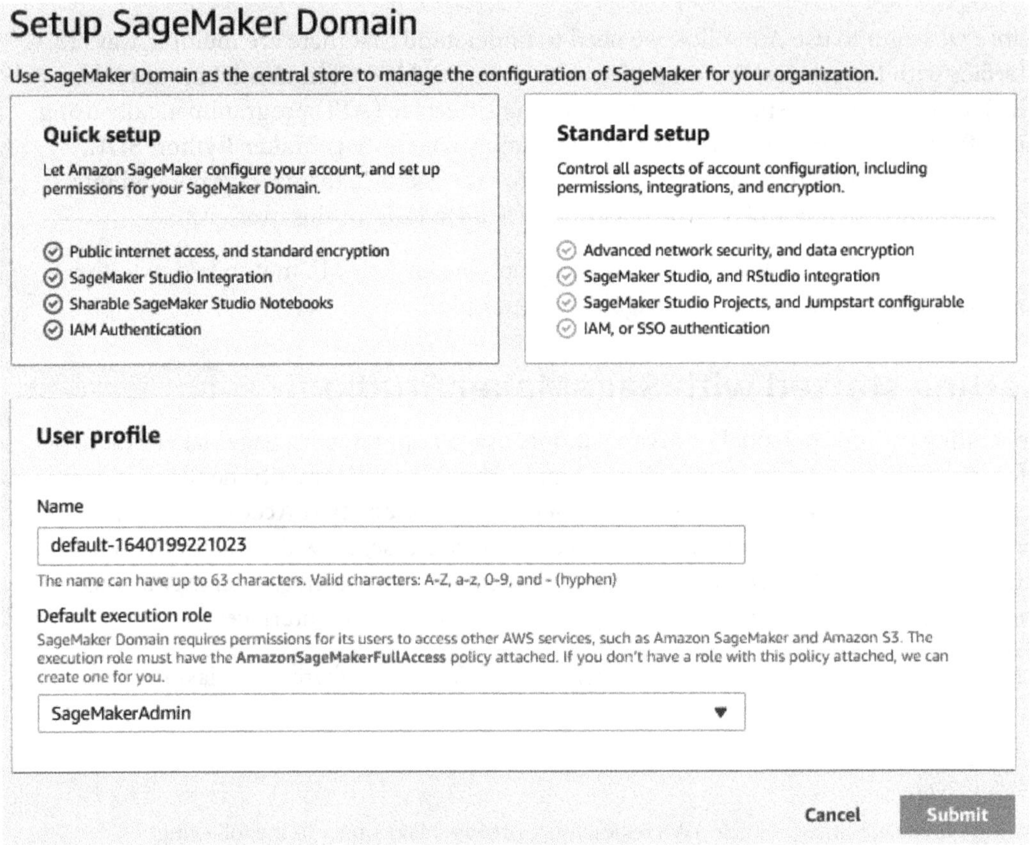

Setup SageMaker Domain

Use SageMaker Domain as the central store to manage the configuration of SageMaker for your organization.

Quick setup	Standard setup
Let Amazon SageMaker configure your account, and set up permissions for your SageMaker Domain.	Control all aspects of account configuration, including permissions, integrations, and encryption.
⊘ Public internet access, and standard encryption	⊘ Advanced network security, and data encryption
⊘ SageMaker Studio Integration	⊘ SageMaker Studio, and RStudio integration
⊘ Sharable SageMaker Studio Notebooks	⊘ SageMaker Studio Projects, and Jumpstart configurable
⊘ IAM Authentication	⊘ IAM, or SSO authentication

User profile

Name

    default-1640199221023

The name can have up to 63 characters. Valid characters: A-Z, a-z, 0-9, and - (hyphen)

Default execution role

SageMaker Domain requires permissions for its users to access other AWS services, such as Amazon SageMaker and Amazon S3. The execution role must have the **AmazonSageMakerFullAccess** policy attached. If you don't have a role with this policy attached, we can create one for you.

    SageMakerAdmin   ▼

                                                         Cancel   Submit

Figure 2.3 – Getting started with SageMaker Studio

5. Select the **Quick setup** option, leaving **Name**, under **User profile**, as the default.

6. Click on the **Default execution role** dropdown and select **Create a new role**.

7. Once the **Create an IAM role** dialog opens, select the **Any S3 Bucket** option, as shown in *Figure 2.4*:

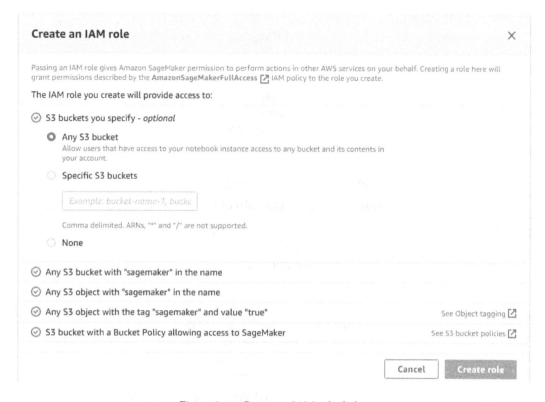

Figure 2.4 – Create an IAM role dialog

8. Click on the **Create role** button to close the dialog box and return to the **Setup SageMaker Domain** screen.

9. Click on the newly created IAM role to open the IAM console summary page dashboard.

10. Now click on the **Add inline policy** link, in the **Permissions policies** section, to open the **Create policy** screen.

11. Click the **Import managed policy** option, at the top right of the **Create policy** screen.

12. Once the **Import managed policies** dialog opens, check the radio button next to **AdministratorAccess**, and then click on the **Import** button.

13. On the **Create policy** screen, click on the **Review policy** button.

14. Once the **Review policy** screen opens, provide a name for the policy, such as `AdminAccess-InlinePolicy`, and then click the **Create policy** button.

> **Note**
>
> Providing administrator access to the SageMaker execution role is not a recommended practice in a production scenario. Since we will access various other AWS services throughout the hands-on examples within this book, we will use the administrator access policy to streamline service permissions.

15. Close the IAM console tab and go back to the SageMaker console.

16. Leave the rest of the **Setup SageMaker Domain** options as their defaults and click the **Submit** button.

17. If you are prompted to select a VPC and subnet, select any subnet in the default VPC and click the **Save and continue** button.

> **Note**
>
> If you are unfamiliar with what a VPC is, you can refer to the following AWS documentation (`https://docs.aws.amazon.com/vpc/latest/userguide/what-is-amazon-vpc.html`).

18. After a few minutes, the SageMaker Studio domain and user will be configured and, as shown in *Figure 2.5*, you should see the SageMaker domain:

Figure 2.5 – Studio control panel

19. Click on the **Launch app** drop-down box next to the default name you created in *step 5* and click **Studio** to launch the Studio IDE web interface.

20. Studio will take a few minutes to launch since this is the first time the Jupyter server is being initialized.

Now that we have the Studio UI online, we can start using Autopilot. But first, we need our raw data.

Preparing the experiment data

Autopilot treats every invocation of the ML process as an experiment and, as you will see, creating an experiment using Studio is simple and straightforward. However, before the experiment can be initiated, we need to provide the experiment with raw data.

Recall from *Chapter 1, Getting Started with Automated Machine Learning on AWS*, that the raw data was downloaded from the UCI repository. We have provided a copy of this data, along with the column names already added in the accompanying GitHub repository (https://github.com/PacktPublishing/Automated-Machine-Learning-on-AWS/blob/main/Chapter02/abalone_with_headers.csv). In order for any of the SageMaker modules to interact with data, the data needs to be uploaded to the AWS cloud and stored as an object, using the Amazon **Simple Storage Service (S3)**.

> **Note**
> You can review the product website (https://aws.amazon.com/s3) if you are unfamiliar with what S3 is and how it works.

Use the following procedure to upload the raw data, for the Autopilot experiment, to S3:

1. Download the preceding file from the accompanying repository to your local machine.

2. To upload the file to Amazon S3, open the S3 console (https://s3.console.aws.amazon.com/s3) in a new web browser tab and then click the bucket name that starts with sagemaker-studio. This S3 bucket was automatically created for you when you used the QuickStart process to onboard to Studio.

3. As *Figure 2.6* shows, the sagemaker-studio-… bucket is empty. To upload the raw data file to the bucket, click the **Upload** button to open the dialog shown in *Figure 2.6*:

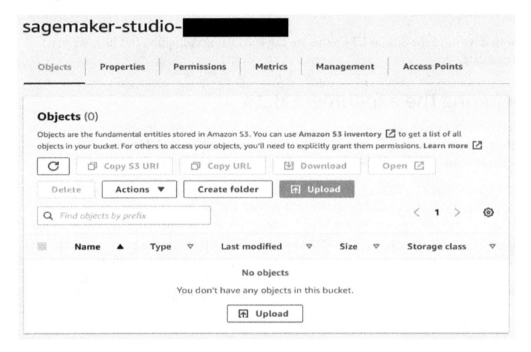

Figure 2.6 – SageMaker Studio bucket

4. On the **Upload** dialog screen, simply drag and drop the abalone_with_ headers.csv file from its download location to the **Upload** dialog screen. Then click the **Upload** button, as shown in *Figure 2.7*:

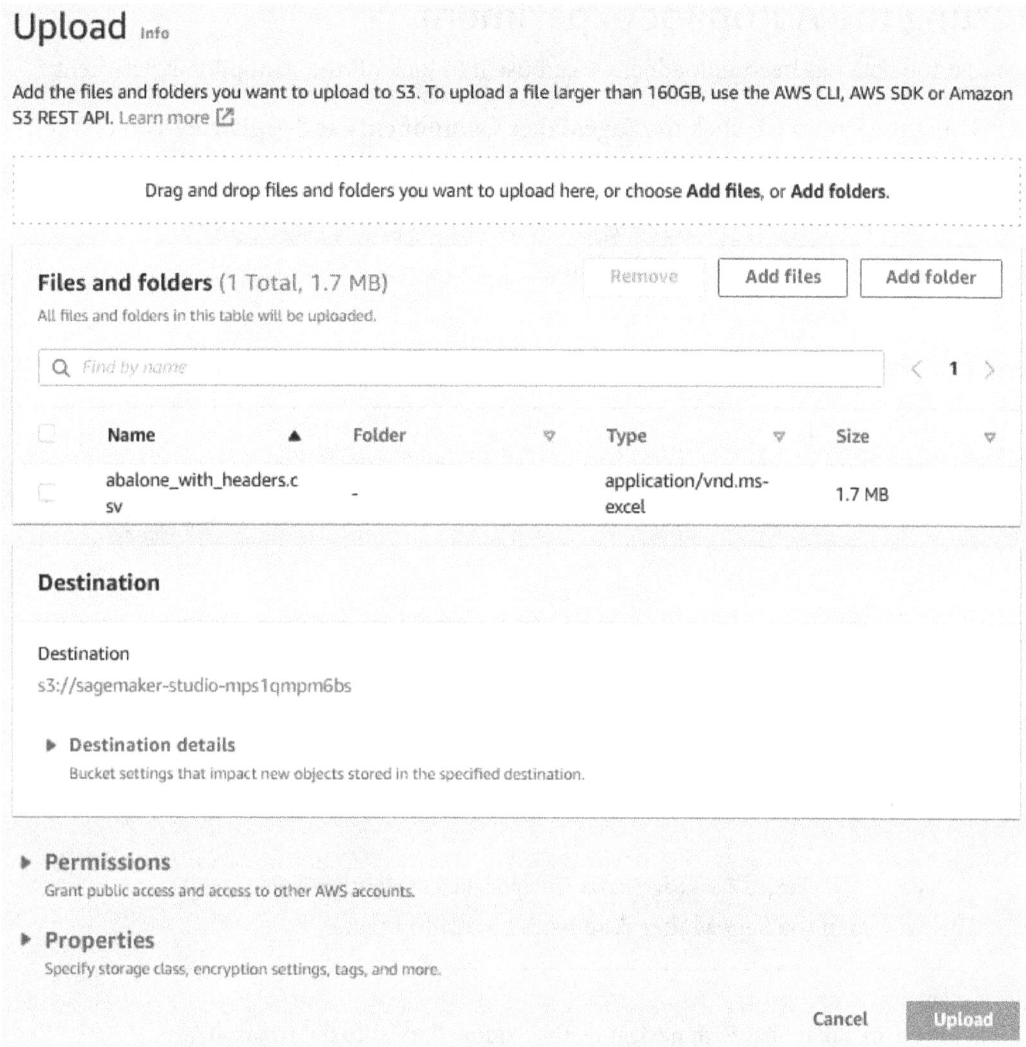

Upload Info

Add the files and folders you want to upload to S3. To upload a file larger than 160GB, use the AWS CLI, AWS SDK or Amazon S3 REST API. Learn more ⤴

Drag and drop files and folders you want to upload here, or choose **Add files**, or **Add folders**.

Files and folders (1 Total, 1.7 MB)   Remove | Add files | Add folder

All files and folders in this table will be uploaded.

	Name ▲	Folder ▽	Type ▽	Size ▽
☐	abalone_with_headers.csv	-	application/vnd.ms-excel	1.7 MB

Destination

Destination

s3://sagemaker-studio-mps1qmpm6bs

▶ **Destination details**
  Bucket settings that impact new objects stored in the specified destination.

▶ **Permissions**
  Grant public access and access to other AWS accounts.

▶ **Properties**
  Specify storage class, encryption settings, tags, and more.

Cancel   Upload

Figure 2.7 – File upload

5.   Once the file has been uploaded, click the **Close** button.

Now we have our data residing in AWS, we can use it to initiate the Autopilot experiment.

Starting the Autopilot experiment

Now that the data has been uploaded, we can use it to kick off the Autopilot experiment:

1. Using the Studio UI, click the **SageMaker Components and registries** icon on the left sidebar:

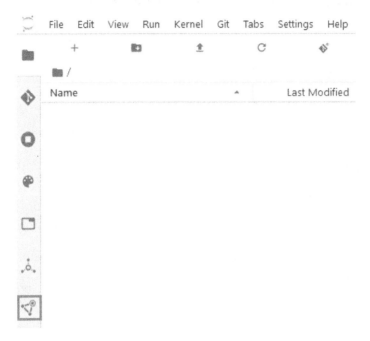

Figure 2.8 – SageMaker Component and registries icon

This will open the **SageMaker resources** navigation pane.

> **Tip**
> If you are unfamiliar with navigating the Studio UI, refer to the Amazon
> SageMaker Studio UI Overview in the AWS documentation (`https://`
> `docs.aws.amazon.com/sagemaker/latest/dg/studio-ui.`
> `html`).

2. From the drop-down menu, select **Experiments and trials** and then click the **Create Autopilot Experiment** button, as shown in *Figure 2.9*, to launch the **Create experiment** tab:

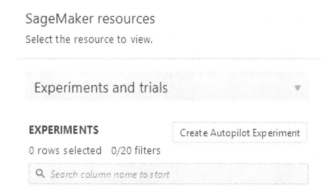

Figure 2.9 – Create Autopilot Experiment

3. The **Create experiment** tab enables you to set the key configuration parameters for the Autopilot experiment.

4. In the **AUTOPILOT EXPERIMENT SETTINGS** dialog, enter the following important settings for the experiment (all other settings can be left at their defaults):

    • **Experiment name:** This is the name of the experiment and it must be unique, in order to track lineage and the various assets created by Autopilot. For this example, enter `abalone-v0` as the experiment name:

AUTOPILOT EXPERIMENT SETTINGS

Figure 2.10 – Experiment name

Tip

In practice, it is a good idea to include the date and time the experiment was initiated or some other form of versioning information, as part of the experiment name. This way, the ML practitioner can easily track the experiment lineage as well as the various experiment assets and ensure these are distinguishable between multiple experiments. For example, if we were creating an experiment for the abalone dataset on July 1, 2021, we could name the experiment `abalone-712021-v0`.

- **Input Data Location**: This is the S3 bucket location that contains the raw dataset you uploaded in the previous section. Using the **S3 bucket name** dropdown, select the bucket that starts with **sagemaker-studio**. Under **Dataset file name**, select the abalone_with_headers.csv file that you previously uploaded:

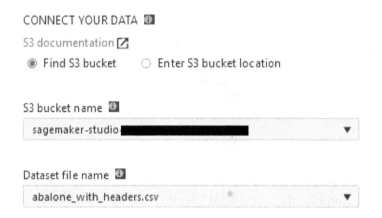

Figure 2.11 – Input data location

- **Target Attribute Name**: This is the name of the feature column, within the raw dataset, on which Autopilot will learn to make accurate predictions. In the **Target** drop-down box, select **rings** as the target attribute:

Figure 2.12 – Target attribute name

> **Note**
> The fact that the ML practitioner must supply a target label highlights a critical factor that must be taken into consideration when using Autopilot. Autopilot only supports supervised learning use cases. Basically, Autopilot will only try to fit supported models for regression and classification (binary and multi-class) problems.

- **Output data location**: The S3 bucket location for any artifacts that are produced by the experiment. From the **S3 bucket name** dropdown, select the bucket that starts with **sagemaker-studio**. Then, enter output for **Dataset directory name** to store the experiment output data:

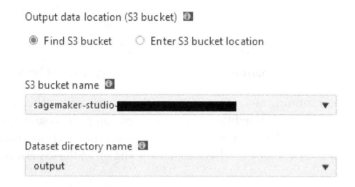

Figure 2.13 – Output data location

- **Problem Type**: This field specifies the type of ML problem to solve. As already noted, this can be a regression, binary classification, or multiclass classification problem. For this example, we will let Autopilot determine which of these problems we are trying to solve by selecting **Auto** from the drop-down box:

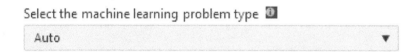

Figure 2.14 – Problem type

> **Tip**
>
> Some of the more experienced ML practitioners might be able to immediately determine the type of problem, based on the use case, and can therefore specify the Autopilot problem type. Should a novice ML practitioner not be able to deduce the type of ML problem, they should set this parameter to **Auto**, since Autopilot has the capability to determine the type of ML problem.

- **Auto deploy**: This option specifies whether to automatically deploy the best model as a production-ready, SageMaker-hosted endpoint. For this example, set the **Auto deploy** option to **Off** so as not to incur unnecessary AWS costs:

Figure 2.15 – Auto deploy

5.   To start the automated experiment, click the **Create Experiment** button.

The experiment is now running and, as you can see, the process of creating a production-grade model using Autopilot is straightforward. However, you are probably wondering what's actually happening in the background, to produce this production-grade model. Let's take a behind-the-scenes look at what's actually going on in the experiment.

Running the Autopilot experiment

Once the experiment has been created, Autopilot will create the best possible candidate for production. The overall process will take approximately 2 hours to complete and the progress can be tracked in a Studio UI tab dedicated to the experiment. *Figure 2.16* shows an example experiment tab:

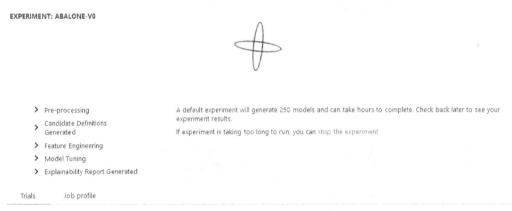

Figure 2.16 – Experiment tab

As the experiment progresses, the various trails that make up the experiment will be displayed in the experiment tab, along with the trial that produces the best model and its overall evaluation score. *Figure 2.17* shows an example of a completed experiment:

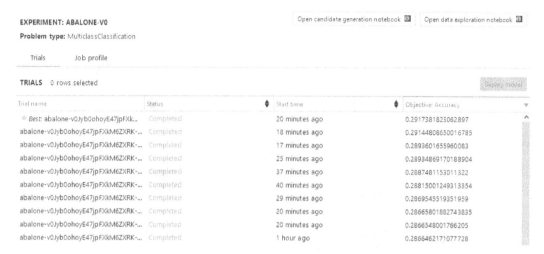

Figure 2.17 – Completed experiment

Once the experiment has been completed, you can right-click on the best model (or any of the other trials) and view the specific details. The following important details on the model are provided:

- The type of ML problem that Autopilot evaluated, based on the raw data

- The algorithm it used to address the assessed ML problem

- The metrics obtained from training the model and used to assess its performance

- The optimization parameters used to tune the model

- The S3 location for the various artifacts that were produced throughout the process

- An explainability report detailing the contribution of each feature, within the raw dataset, to the prediction

- The capability to deploy the model as a SageMaker hosted endpoint

So, by means of a simple process, all the heavy lifting tasks for data analysis, model building, training, evaluation, and tuning have been automated and managed by Autopilot, making it easy for the novice ML practitioner to overcome the two main challenges imposed by the ML process.

While this may suffice for an inexperienced application developer or novice ML practitioner to simply get a model into production, a more experienced ML practitioner may require more proof of why the particular model is the best and how it was produced. *Figure 2.18* shows an overview of how Autopilot produces the best models:

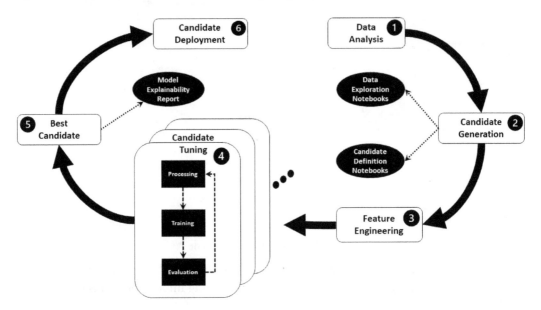

Figure 2.18 – Overview of the AutoML process used by Autopilot

As you can see from *Figure 2.18*, there are six key tasks that Autopilot is automatically executing in the background:

1. **Data Analysis**
2. **Candidate Generation**
3. **Feature Engineering**
4. **Candidate Tuning**
5. **Best Candidate**
6. **Candidate Deployment**

Let's follow each step of the process in detail.

Data preprocessing

The first step of the ML process is to access the raw dataset and understand it, in order to clean it up and prepare it for model training. Autopilot does this automatically for us executing the data analysis and preprocessing step. Here, Autopilot leverages the SageMaker Processing module to statistically analyze the raw dataset and determine whether there are any missing values. Autopilot then shuffles and splits the data for model training and stores the output data in S3. Once the raw data has been preprocessed, it's ready for the model candidate generation step.

Should Autopilot encounter any missing values within the dataset, it will attempt to fill in the missing data using a number of different techniques. For example, for any missing categorical values, Autopilot will create a distinct *unknown* category feature. Alternatively, for any missing numeric values, Autopilot will try to impute the value using the *mean* or *median* of the feature column.

Generating AutoML candidates

The next step that Autopilot performs is to generate model candidates. In essence, each candidate is an AutoML pipeline definition that details the individual parts of a workflow that produces an optimized model candidate, or best model. Based on Autopilot's statistical understanding of the data, each candidate definition details the type of model to be trained and then, based on that model candidate, the data transformations necessary to engineer features that best suit the algorithm.

Depending on which *problem type* setting was specified when creating the experiment, Autopilot will select the appropriate algorithm from SageMaker's built-in estimators. In the case of the *Abalone Calculator* example, **Auto** was selected, and therefore, Autopilot deduces from analyzing the dataset that this is a regression problem, so it creates candidate definitions that each implement a variation of the *Linear Learner Algorithm*, *XGBoost Algorithm*, and a *Multi-layer Perceptron* deep learning algorithm. Each of these candidates has its own set of training and testing data, as well as the specific ranges of hyperparameters to tune on. Autopilot creates up to 10 candidate definitions.

Tip

For more information on Autopilot's supported algorithms, you can review the Model Support and Validation section of the AWS documentation (`https://docs.aws.amazon.com/sagemaker/latest/dg/autopilot-model-support-validation.html`).

Figure 2.18 highlights the outputs from the **Candidate Generation** step, two Jupyter notebooks:

- *Data Exploration Notebooks*: This notebook provides an overview of the data that was analyzed during the **Data Analysis** step and provides guidance for the ML practitioner to investigate further.

- *Candidate Definition Notebooks*: This notebook provides a detailed overview of the model candidates, recommended data processing for the candidate, and what hyperparameters should be tuned to optimize the candidate. The notebook even generates Python code cells, with the appropriate SageMaker SDK calls, to reproduce the candidate pipeline, thus giving the novice ML practitioner a how-to guide on reproducing a production-ready model.

If you recall from *Chapter 1, Getting Started with Automated Machine Learning on AWS*, one of the earmarks of an efficient AutoML process is the fact that the process must be repeatable. By providing candidate definition notebooks, not only does Autopilot provide a how-to guide for the ML practitioner but also allows them to build upon the process and create their own candidate pipelines.

> **Tip**
> Since AutoML technically only needs to be executed once, to get the best candidate for production deployment, these notebooks can be used as a foundation to further customize and develop the model.

Before these model candidates can be trained, the raw data must be formatted to suit the specific algorithm that the candidate pipeline will use. This process happens next.

Automated feature engineering

The next step of the AutoML process is the feature engineering stage. Here, Autopilot once again leverages the SageMaker Processing module to engineer these new features, specific to each model candidate. Autopilot then creates training and validation dataset variations that include these features and stores these on S3. Each candidate now has its own formatted training and testing dataset. Now the training process can begin.

Automated model tuning

At this stage of the process, Autopilot has the necessary components to train each of the candidate models. Unlike the *typical* ML process, where each candidate is trained, tuned, and evaluated, Autopilot leverages SageMaker's automatic model tuning module to execute the process in parallel.

As explained at the outset of this chapter, the hyperparameter optimization module uses *Bayesian Search* to find the best parameters for the model. However, Autopilot takes this one step further and leverages SageMaker's native capability to extend the tuning capability across multiple algorithms as well. In essence, Autopilot not only finds the best hyperparameters for an individual model candidate but the best hyperparameters when compared to all the other model candidates.

As already mentioned, Autopilot performs this process in parallel, training, tuning, and evaluating each model candidate with a subset of hyperparameters in order to get the best model candidate and associated hyperparameters for that subset, as a trial. The process is then repeated, constantly refining the hyperparameters, up to the default of 250 trials. This capability greatly reduces the overall time taken to produce an optimized model to a matter of hours as opposed to days or weeks when using a manual ML process.

The tuning process produces up to 250 candidate models. Let's review these candidate models next.

Candidate model selection

As *Figure 2.17* highlights, the model that produces the best evaluation metric result is labeled **Best**.

The outputs from this step are the models and the associated artifacts for each trial and an explainability report. Autopilot uses another SageMaker module, called **SageMaker Clarify**, to produce this report.

Clarify helps ML practitioners understand how and why trained models make certain predictions, by quantifying the contribution that each feature of the dataset makes towards the model's overall prediction. This helps not only the ML practitioner but also the use case stakeholders, to understand how the model determines its predictions. Understanding why a model makes certain predictions promotes further trust in the model's capability to address the goals and requirements of the business use case.

> **Tip**
>
> For more information on the process that SageMaker Clarify uses to quantify feature attributions, you can refer to the Model Explainability page in the SageMaker documentation (`https://docs.aws.amazon.com/sagemaker/latest/dg/clarify-model-explainability.html`).

The experiment is now complete, so all that's left for the ML practitioner to do is deploy the *best* candidate into production.

Candidate model deployment

The best model can now be automatically, or manually, deployed as a **SageMaker Hosted Endpoint**. This means that the best candidate is now an API that can be programmatically called upon to provide predictions for the production application. However, simply deploying the model into production doesn't stop the ML process. The process is continuous and there are a number of tasks still to be performed after experimentation.

Post-experimentation tasks

In the previous chapter, we emphasized that the **CRISP-DM** methodology ends with the model being deployed into production, and we also highlighted that producing a production-ready model is not necessarily the conclusion of an ML practitioner's responsibilities.

The same concepts apply to the AutoML process. While Autopilot takes care of the various steps to generate a production-ready model, this is typically a one-time process for the specific use case and Autopilot concludes the experiment after the model has been deployed. On the other hand, the ML practitioner's obligations are ongoing since the production model needs to be continuously monitored to ensure that it does not drift from its intended purpose.

However, by providing all the output artifacts for each model candidate as well as the candidate notebooks, Autopilot lays a firm foundation for the novice ML practitioner to close the loop on the ML process and continuously optimize future production models, should the deployed model drift from its intended purpose.

Additionally, SageMaker hosted endpoints provide added functionality to assist with the process of continuously monitoring the production model for concept drift. For example, should the ML practitioner decide to manually deploy the best candidate model in the Studio UI, they can enable data capture when selecting the best candidate and clicking on the **Deploy Model** button. *Figure 2.19* shows the options available before deploying the model using the Studio UI:

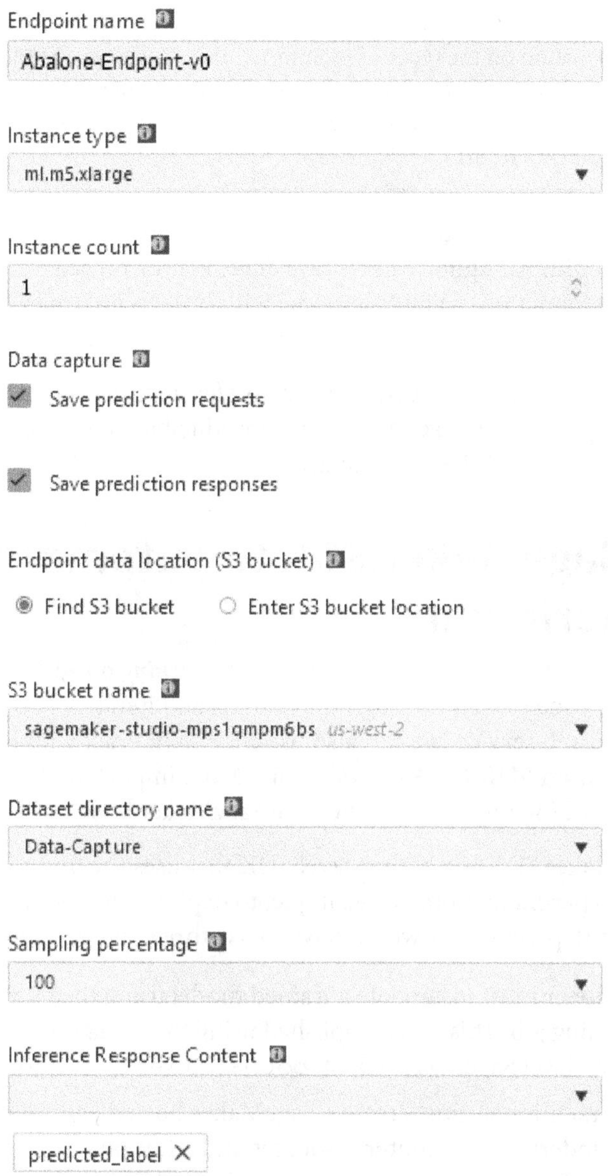

Endpoint name ⓘ

Abalone-Endpoint-v0

Instance type ⓘ

ml.m5.xlarge ▾

Instance count ⓘ

1 ⌄

Data capture ⓘ

☑ Save prediction requests

☑ Save prediction responses

Endpoint data location (S3 bucket) ⓘ

◉ Find S3 bucket   ○ Enter S3 bucket location

S3 bucket name ⓘ

sagemaker-studio-mps1qmpm6bs *us-west-2* ▾

Dataset directory name ⓘ

Data-Capture ▾

Sampling percentage ⓘ

100 ▾

Inference Response Content ⓘ

▾

predicted_label ✕

Figure 2.19 – Deployment options

Enabling the **Data capture** settings configures the endpoint to capture all incoming requests for prediction as well as the prediction responses from the deployed model. The captured data can be used by SageMaker's Model Monitor feature to monitor the production model in real time, continuously assessing its performance on unseen requests and looking for concept deviations.

> **Tip**
>
> For more information on the types of monitoring that SageMaker Model Monitor performs, you can refer to the SageMaker documentation (`https://docs.aws.amazon.com/sagemaker/latest/dg/model-monitor.html`).

From this example, you can see that by utilizing the Studio UI to create and manage an AutoML experiment (with Autopilot), a software developer or novice ML practitioner can easily produce a production-grade ML model with little to no experience and without writing any code.

However, some experienced ML practitioners may prefer documents and codify the experiment using a Jupyter notebook so that it is reproducible. In the next section, we will look at how to codify the AutoML experiment.

Using the SageMaker SDK to automate the ML experiment

In *Chapter 1*, *Getting Started with Automated Machine Learning on AWS*, you were provided with sample code to walk through the manual and iterative ML process. Since SageMaker is an AWS web service, we can also use code to interact with its various modules using the **Python SDK for AWS**, or **boto3**. More importantly, AWS also provides a dedicated Python SDK for SageMaker, called the SageMaker SDK.

In essence, the SageMaker SDK is a higher-level SDK that uses the underlying boto3 SDK with a focus on ML experimentation. For example, to deploy a model as a SageMaker hosted endpoint, an ML practitioner would have to use three different boto3 calls:

1. The ML practitioner must instantiate a trained model using the output artifact from a SageMaker training job. This is accomplished using the `create_model()` method from boto3's low-level SageMaker client, `boto3.client("sagemaker")`.

2. Next, the ML practitioner must create a SageMaker hosted endpoint configuration, specifying the underlying computer resources and additional configuration settings for the endpoint. This is done using the `create_endpoint_config()` method from the SageMaker client.

3. Finally, the `create_endpoint()` method is used to deploy the trained model with the endpoint configuration settings.

Alternatively, the ML practitioner could accomplish the same objective by simply using the `deploy()` method on an already trained model, using the SageMaker SDK. The SageMaker SDK creates the underlying model, endpoint configuration, and automatically deploys the endpoint.

Using the SageMaker SDK makes ML experimentation much easier for the more experienced ML practitioner. In this next section, you will start familiarizing yourself with the SageMaker SDK by working through an example to codify the AutoML experiment.

Codifying the Autopilot experiment

In the same way that an ML practitioner uses a Jupyter notebook to execute a manual and interactive ML experiment, the Studio UI can also be used to accomplish these same tasks. The Studio IDE provides basic Jupyter Notebook functionality and comes pre-installed with all the Python libraries and deep learning frameworks an ML practitioner might use, in the form of AWS engineered Jupyter kernels.

> Tip
> If you are unfamiliar with the concept of a Jupyter Kernel and how they are used, you can refer to the Jupyter documentation website (`https://jupyter-notebook-beginner-guide.readthedocs.io/en/latest/what_is_jupyter.html#kernel`).

Let's see how this works by using a Jupyter notebook to execute the following sample code:

1. Using the Studio UI menu bar, select **File | New | Notebook** to open a blank Jupyter notebook.

2. As *Figure 2.20* shows, you will be prompted to select an appropriate kernel from the selection of pre-installed kernels. From the drop-down list, select **Python 3 (Data Science)**:

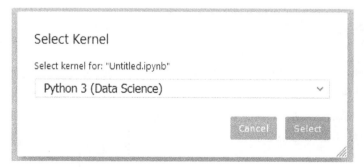

Figure 2.20 – Jupyter kernel selection

3. In the background, Studio will initialize a dedicated compute environment, called a **KernelGateway**. This compute environment, with *2 x vCPUs and 4 GB of RAM*, is in essence the engine that executes the various code cells within the Jupyter notebook. It may take 2–3 minutes to initialize this compute environment.

4. Once the Kernel has started, we can create the first code cell, where we import the SageMaker SDK and configure the SageMaker session by initializing the `Session()` class. The `Session()` class is a wrapper for the underlying boto3 client, which governs all interactions with the SageMaker API, as well as other necessary AWS services:

```
import sagemaker
import pandas as pd
role = sagemaker.get_execution_role()
session = sagemaker.session.Session()
```

> **Tip**
>
> If you are new to navigating through a Jupyter notebook, to execute a code cell, you can either click on the run icon or press the *Shift + Enter* keys on the highlighted cell.

5. Next, we can use the same code we used in *Chapter 1, Getting Started with Automated Machine Learning on AWS*, to download the raw abalone dataset from the UCI repository, add the necessary column headings, and then save the file as a CSV file called `abalone_with_headers.csv`:

```
column_names = ["sex", "length", "diameter", "height",
"whole_weight", "shucked_weight", "viscera_weight",
"shell_weight", "rings"]
abalone_data = pd.read_csv("http://archive.ics.uci.edu/
ml/machine-learning-databases/abalone/abalone.data",
names=column_names)
abalone_data.to_csv("abalone_with_headers.csv",
index=False)
```

6. Now, we can configure the Autopilot experiment, using the `AutoML()` class and using the following example code to create a variable called `automl_job`:

```
from sagemaker.automl.automl import AutoML
automl_job = AutoML(
    role=role,
    target_attribute_name="rings",
```

```
    output_path=f"s3://{session.default_bucket()}/
abalone-v1/output",
    base_job_name="abalone",
    sagemaker_session=session,
    max_candidates=250
)
```

This variable will interact with the Autopilot experiment. As we did in the previous section, we also supply the important parameters for the experiment:

- `target_attribute_name`: The name of the feature column within the raw dataset on which Autopilot will learn to make accurate predictions. Recall that this is the `rings` attribute.

- `output_path`: Where any artifacts that are produced by the experiment are stored in S3. In the previous example, we used the default S3 bucket that was created during the Studio onboarding process. However, in this example, we will use the default bucket provided by the SageMaker `Session()` class.

- `base_job_name` or the name of the Autopilot experiment. Since we already have a version zero, we will name this experiment `abalone-v1`.

> **Note**
>
> No versioning information has been supplied to the `base_job_name` parameter. This is because the SDK automatically binds the current date and time for versioning.

7. To initiate the Autopilot experiment, we call the `fit()` method on the `automl_job` variable, supplying it with the S3 location of the raw training data. We also call the SageMaker session `upload()` method, as a parameter to the `fit()` method since the data has not yet been uploaded to the default S3 bucket. The `upload()` method takes the bucket name (the SageMaker default bucket) and the prefix (the folder structure) as parameters to automatically upload the raw data to S3. The following code shows an example of how to correctly call the `fit()` method:

```
automl_job.fit(inputs=session.upload_data("abalone_with_
headers.csv", bucket=session.default_bucket(), key_
prefix="abalone-v1/input"), wait=False)
```

At this point, the Autopilot experiment has been started and, as was shown in the previous section, the experiment can be monitored using the **Experiments and trials** dropdown in the **SageMaker Components and registries** section of the Studio UI. Right-click on the current Autopilot experiment and select **Describe AutoML Job**.

Alternatively, we can use the `describe_auto_ml_job()` method on the `automl_job` variable to programmatically get the current overview of the Autopilot job.

> **Note**
>
> Make a note of the automatically generated versioning information that the SDK appends to the job name. As you will see later, this job name is used to programmatically explore the experiment as well as to clean it up.

To make the experiment reproducible, some ML practitioners might want to include a visual comparison of the resultant models. Now that the AutoML experiment is underway, we can wait for it to complete to see how the models compare and explore how the SageMaker SDK enables experiment analysis.

Analyzing the Autopilot experiment with code

Once the Autopilot experiment has been completed, we can use the `analytics()` class from the SageMaker SDK to programmatically explore the various model candidates (or trials) to compare candidate evaluation results, in the same way we used the Studio UI in the previous section.

Let's analyze the experiment by using the same Jupyter notebook to execute the following sample code:

1. The first thing we need to do is load the `ExperimentAnalytics()` class from the SageMaker SDK to get the trial component data and make them available for analysis. By providing the name of the Autopilot experiment, the following sample code instantiates the `automl_experiment` variable, whereby we can interact with the experiment results. Additionally, since the SageMaker SDK automatically generates the versioning information for the experiment name, we can once again use the `describe_auto_ml_job()` method to find the **AutoMLJobName**:

    ```
    from sagemaker.analytics import ExperimentAnalytics
    automl_experiment = ExperimentAnalytics(
        sagemaker_session=session,
        experiment_name="{}-aws-auto-ml-job".format(automl_
    job.describe_auto_ml_job()["AutoMLJobName"])
    )
    ```

2. Next, the following sample code converts the returned `experiment` analytics object into a pandas DataFrame for easier analysis:

    ```
    df = automl_experiment.dataframe()
    ```

3. An example of such analysis might be to visually compare the evaluation results of the top five trials. The following sample code filters the DataFrame by the latest evaluation accuracy metrics, `validation:accuracy - Last` and `train:accuracy - Last`, on both the training dataset and validation dataset respectively, and then sorts these values in ascending order:

```
df = df.filter(["TrialComponentName","validation:accuracy
- Last", "train:accuracy - Last"])
df = df.sort_values(by="validation:accuracy - Last",
ascending=False)[:5]
df
```

Figure 2.21 shows what an example of this DataFrame would look like:

	TrialComponentName	validation:accuracy - Last	train:accuracy - Last
5	abalone-2021-07-05-17-07-15-99Qg-240-9bfa065e-...	0.292639	0.267620
58	abalone-2021-07-05-17-07-15-99Qg-192-2481d808-...	0.289939	0.275474
155	abalone-2021-07-05-17-07-15-99Qg-097-993860a4-...	0.288151	0.281461
142	abalone-2021-07-05-17-07-15-99Qg-109-7b9e8357-...	0.287852	0.285575
16	abalone-2021-07-05-17-07-15-99Qg-237-27a447ff-...	0.286952	0.281460

Figure 2.21 – Sample top 5 trials

4. We can further visualize the comparison using a plot by means of the `matplotlib` library:

```
import matplotlib.pyplot as plt
%matplotlib inline

legend_colors = ["r", "b", "g", "c", "m"]
ig, ax = plt.subplots(figsize=(15, 10))
legend = []
i = 0
for column, value in df.iterrows():
    ax.plot(value["train:accuracy - Last"],
value["validation:accuracy - Last"], "o", c=legend_
colors[i], label=value.TrialComponentName)
    i +=1
plt.title("Training vs.Testing Accuracy",
fontweight="bold", fontsize=14)
```

```
plt.ylabel("validation:accuracy - Last",
fontweight="bold", fontsize=14)
```

```
plt.xlabel("train:accuracy - Last", fontweight="bold",
fontsize=14)
```

```
plt.grid()
```

```
plt.legend()
```

```
plt.show()
```

Figure 2.22 shows an example of the resultant plot:

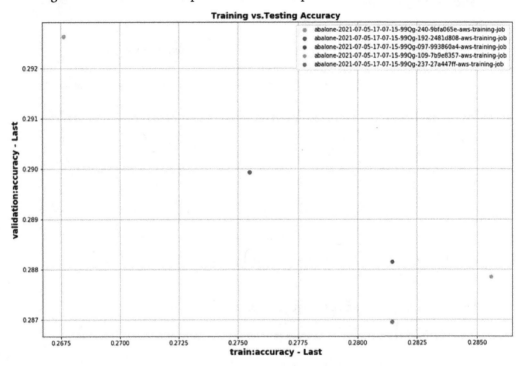

Figure 2.22 – Plot of the top 5 trials

Using the `ExperimentAnalytics()` class is a great way to interact with the various trials of the experiment, however, you want to simply see which trial produces the best candidate. By calling the `best_candidate()` method on the Autopilot job, we can not only see which trial produced the best candidate, but also the value of the candidate's final evaluation metric. For example, the following sample code produces the name of the best candidate:

```
automl_job.best_candidate()["CandidateName"]
```

When executing the preceding code in the Jupyter notebook, you will see an output similar to the following:

```
'abalone-2021-07-05-17-07-15-99Qg-240-9bfa065e'
```

Likewise, the following sample code can be executed in an additional notebook cell to see the best candidate's evaluation metrics:

```
automl_job.best_candidate()["FinalAutoMLJobObjectiveMetric"]
```

The results of this code will be similar to the following:

```
{'MetricName': 'validation:accuracy', 'Value':
0.292638897895813}
```

Additionally, just as with the example in the previous section, you can also programmatically view the S3 location of the **Data Exploration notebook**, **Candidate Definition notebook**, and the **Explainability Report**.

The following code samples can be used to get this information:

- **Data Exploration Notebook**:

  ```
  automl_job.describe_auto_ml_job()["AutoMLJobArtifacts"]
  ["DataExplorationNotebookLocation"]
  ```

- **Candidate Definition Notebook**:

  ```
  automl_job.describe_auto_ml_job()["AutoMLJobArtifacts"]
  ["CandidateDefinitionNotebookLocation"]
  ```

- **Explainability Report**:

  ```
  automl_job.describe_auto_ml_job()["BestCandidate"]
  ["CandidateProperties"]["CandidateArtifactLocations"]
  ["Explainability"]
  ```

Using the `analytics()` class of the SageMaker SDK and, the various Autopilot output artifacts has allowed us to gain further insight into the experiment. All that's left is to deploy the production model.

Deploying the best candidate

The last part of the AutoML process is to deploy the best model as a production API, as a SageMaker hosted endpoint. To provide for this functionality, the SageMaker SDK once again provides a simple method, called `deploy()`.

> **Note**
>
> Hosting the best model on SageMaker will incur AWS usage costs that exceed what is provided by the free tier.

Let's run the following code to deploy the best model:

```
automl_job.deploy(
    initial_instance_count=1,
    instance_type="ml.m5.xlarge",
    candidate=automl_job.best_candidate(),
    sagemaker_session=session,
    endpoint_name="-".join(automl_job.best_candidate()
["CandidateName"].split("-")[0:7])
)
```

As was the case with the `fit()` method, simply calling the `deploy()` method and providing some important parameters will create a hosted endpoint:

- We need to supply the type of computer resources to process inference requests by supplying the `instance_type` parameter. In this case, we selected the `ml.m5.xlarge` instance.

- We then need to specify the number of instances. In this case, we are specifying one instance.

- We can either deploy a specific candidate, by providing the Python dictionary for the specific candidate or, if no candidate is provided, the `deploy()` method will automatically use the best candidate.

- Lastly, we need to provide the unique name for the endpoint. This name will be used to provide model predictions to the business application.

> **Tip**
>
> When naming the endpoint, it is a good practice to use the versioning information supplied to the experiment in order to tie specific endpoints to the experiment that produced them. From the preceding sample Python code, you can see that we derived `endpoint_name` by using the `best_candidate()` method and filtering the response by `"CandidateName"`.

Once the code cell is executed, SageMaker will automatically deploy the best model on the specific compute resources and make the endpoint API available for inference requests, thus completing the experiment.

> **Note**
>
> Unlike the preceding example, where the Studio UI is used to deploy the model, using the `AutoML()` class from the SageMaker SDK does not include the ability to enable data capture when deploying a model. Recall that the ability to capture both inference requests and inference responses enables the ML practitioner to use this data to continuously monitor a production model. It is recommended that you use the SageMaker SDK's `Model()` class to deploy the model. This class allows you to specify the `data_capture_config` parameter, should you wish to close the loop on continuous model monitoring. You can learn more about the `Model()` class in the SageMaker SDK documentation (`https://sagemaker.readthedocs.io/en/stable/api/inference/model.html#sagemaker.model.Model.deploy`).

As was the case with the *typical* ML process, highlighted in *Chapter 1*, *Getting Started with Automated Machine Learning on AWS*, the deployed model can now be handed over to the application development owners for them to test and integrate the model into the production application. However, since the intended purpose of this example was to simply demonstrate how to make the required SageMaker SDK calls to execute the AutoML experiment, we are not going to use the model to test inferences. Instead, the next section demonstrates how to delete the endpoint.

Cleaning up

To avoid unnecessary AWS usage costs, you should delete the SageMaker hosted endpoint. This can be accomplished by using the AWS SageMaker console (https://console. aws.amazon.com/sagemaker) or by using the AWS CLI. Run the following commands in the Jupyter notebook to clean up the deployment:

1. Using the AWS CLI, delete the SageMaker hosted endpoint:

```
!aws sagemaker delete-endpoint --endpoint-name {"-".
join(automl_job.best_candidate()["CandidateName"].
split("-")[0:7])}
```

2. Then use the AWS CLI to also delete the endpoint configuration:

```
!aws sagemaker delete-endpoint-config --endpoint-
config-name {"-".join(automl_job.best_candidate()
["CandidateName"].split("-")[0:7])}
```

> **Tip**
> If you wish to further clean up the various trials from the experiment,
> you can refer to the Clean Up section of the SageMaker documentation
> (https://docs.aws.amazon.com/sagemaker/latest/dg/
> experiments-cleanup.html).

From the perspective of automating the ML process, you should now be acquainted with how the Autopilot module can be used to realize an AutoML methodology, and how the SageMaker SDK can be used to create a codified and documented AutoML experiment.

Summary

This chapter introduced you to some of AWS's AI and ML capabilities, specifically Amazon SageMaker's. You saw how to interact with the service via the SageMaker Studio UI and the SageMaker SDK. Using hands-on examples, you learned how Autopilot's implementation of the AutoML methodology addresses not only the two challenges imposed by the *typical* ML process but also the overall criteria for automation. Particularly, how using Autopilot ensures that the ML process is *reliable* and *streamlined*. The only task required to be done by the ML practitioner is to upload the raw data to Amazon S3.

This chapter also highlights an important aspect of the AutoML methodology. While the AutoML process is *repeatable* in the sense that it will always produce an optimized model, once you have the model in production, there is no real need to recreate it, unless, of course, the business use case changes. Nevertheless, Autopilot creates a solid foundation to help an ML practitioner continuously optimize the production model, by providing **Candidate Notebooks**, model **Explainability Reports**, and enabling **Data Capture** to monitor the model for concept drift. So, using the AutoML methodology is a great way for ML practitioners to automate their initial ML experiments.

Worth noting is a drawback of Autopilot's implementation of the AutoML methodology. Autopilot only supports **Supervised Learning** use cases – ones that use **Regression** or **Classification** models.

In the next chapter, you will learn how to apply the AutoML methodology to more complicated ML use cases that require advanced deep learning models, using the AutoGluon package.

3
Automating Complicated Model Development with AutoGluon

In *Chapter 1*, *Getting Started with Automated Machine Learning on AWS*, you were introduced to the *ACME Fishing Logistics* use case, where you created a production-grade MLP model using a typical ML process. While the example only highlights a basic artificial neural network architecture, it also provides a suitable introduction to the concept of deep learning.

Deep learning is an advanced ML technique that can be used to solve complex and challenging use cases such as customer sentiment analysis, language translation, and object detection images and videos. These complex use cases often require the ML practitioner to create very intricate, as well as exceptionally large, neural network architectures. Some of these architectures can have hundreds of thousands, even billions, of trainable parameters. The more complicated the network, the more challenging it becomes to train and therefore, the more challenging it becomes to automate.

As we highlighted in the previous chapter, SageMaker Autopilot only supports tabular data. Thus, more complicated deep learning use cases that would require image data are not supported. *So how can we apply an AutoML methodology to automate the ML process for these complex use cases, ones that require deep learning models?*

In this chapter, we will investigate how this can be accomplished by using the AutoGluon Python library and illustrate how Amazon SageMaker can still be used to effectively apply an AutoML methodology to some of these more complex deep learning use cases. We will also use this opportunity to show you some of the cutting-edge capabilities of SageMaker by introducing you to SageMaker's **Bring Your Own Container** (**BYOC**) functionality, as well as AWS **Deep Learning Containers**. We will be making use of this capability quite extensively throughout the book. Specifically, we will focus on the following topics:

- Introducing the AutoGluon library
- Using AutoGluon for tabular data
- Using AutoGluon for image data

By the end of the chapter, you will have a practical understanding of what the AutoGluon library is and how to use it.

Technical requirements

Since we will once again be using the SageMaker Studio UI, the technical requirements for this chapter are the same as *Chapter 2, Automating Machine Learning Model Development Using SageMaker Autopilot*:

- A web browser (for the best experience, it is recommended that you use a Chrome or Firefox browser).
- Access to the AWS account that you used in *Chapter 2, Automating Machine Learning Model Development Using SageMaker Autopilot*.
- We will once again be working within the usage limits of the AWS Free Tier to avoid exceeding unnecessary costs.

Introducing the AutoGluon library

AutoGluon is a Python library developed by AWS and open sourced at their annual *re:Invent conference*, in 2019. The primary design goal behind AutoGluon is similar to SageMaker's Autopilot module – to resolve all the complexities and challenges an ML practitioner faces in a *typical* ML process and resolve these with a single Python library. In essence, AutoGluon empowers the ML practitioner to organize their training data and apply several ML approaches to generate an optimized model, all with just a few lines of code.

AutoGluon overcomes some of the limitations that AutoPilot has in that it can address the more complex ML use cases that involve compound types of data, such as cluttered text data and images. Of course, AutoGluon also works with tabular data. AutoGluon accomplishes this by creating separate predictors for each data type and, hence, each type of ML use case that the data type supports. For example, AutoGluon includes the following predictors:

- **Tabular Predictor**: This predictor is like Autopilot's functionality as it is used to create an optimized model to predict column values from tabular data and, just like Autopilot, this applies to both the classification and regression use cases.

- **Image Predictor**: This predictor focuses on generating models that predict the named category for entire images. For example, if we had a dataset containing labeled images of cats and dogs, the image predictor would create an optimized model to predict whether a new image falls under the *cat* or *dog* category.

- **Object Detector**: Using this predictor builds an optimized model with the ability to distinctly recognize different objects in a single image. For example, if we supply a model trained using the object detector with an image of a *boy* and his *dog*, the model would be able to differentiate between the two individual objects.

- **Text Predictor**: This predictor provides functionality similar to the Tabular Predictor in that it creates an optimized model to perform regression and classification prediction tasks on text data. For example, if a model were optimized using the Text Predictor, given a string of text, it would be able to classify the sentiment of the sentence.

> **Note**
>
> The Text Predictor uses tabular training data for classification and regression in a similar way to how the Tabular Predictor uses tabular data. The key difference is that the Tabular Predictor will feature the text while the Text Predictor will fit directly to the raw text. In other words, the Tabular Predictor will convert the text columns of the tabular data into a vector (or numerical) representation. On the other hand, the Text Predictor will work directly with the raw text.

While this is just a high-level introduction to AutoGluon, the best way to see its capabilities in action is to work through a hands-on example. Let's experiment with AutoGluon on the *ACME Fishing Logistics* use case.

> **Tip**
>
> For more information on these built-in predictors and how to make a single Python call to fit these onto the raw data, you can review the AutoGluon documentation (`https://auto.gluon.ai/stable/api/autogluon.predictor.html`).

Using AutoGluon for tabular data

In the previous chapter, we used Autopilot to see an example AutoML experiment that applies to the *ACME Fishing Logistics* use case. In this example, we are going to reproduce this experiment with AutoGluon. So, let's see how we can use AutoGluon to automate this task.

> **Note**
>
> The AutoGluon Tabular library benefits from compute instances with as much memory as possible. It is, therefore, recommended that AWS M5 instances (`https://aws.amazon.com/ec2/instance-types/m5/`) are used for tabular experiments. We will be using an `m5.xlarge` instance in this example and, therefore, running the example will incur AWS resource costs.

Prerequisites

Before we begin, there are a few fundamental topics that need to be accounted for, namely:

- At the time of writing, the AutoGluon library is not natively included as one of SageMaker's built-in estimators. This means that we will have to create our own Docker container for AutoGluon, using the SageMaker BYOC methodology and AWS Deep Learning Containers.

- Unless we use SageMaker managed notebook instances (`https://docs.aws.amazon.com/sagemaker/latest/dg/nbi.html`), which have native support for the Docker daemon, there is no inherent functionality for building Docker containers when using SageMaker Studio.

To address these constraints while still using the Studio UI for this example, AWS has provided an open source CLI utility called `sagemaker-studio-image-build` (`https://github.com/aws-samples/sagemaker-studio-image-build-cli`). This utility allows us to build a SageMaker-compatible container within the Studio UI and, that way, we can now build and run the AutoGluon example from within the Studio environment. In the background, the `sagemaker-studio-image-build` library uses the fully managed build service, AWS CodeBuild, to build the Docker image. To access the service, the Studio execution role requires the appropriate access.

> **Note**
>
> If you are unfamiliar with AWS CodeBuild and how it works, you can refer to the product web page (`https://aws.amazon.com/codebuild`).

Configuring service permissions

The following steps will walk you through how to configure the appropriate permissions for the SageMaker execution role:

1. Log in to your AWS account, navigate to the **Amazon SageMaker** management console, and click the **SageMaker Domain** link in the left-hand navigation panel.

> **Note**
>
> You should already have onboarded to SageMaker Studio. If not, refer to the *Getting started with SageMaker Studio* section in *Chapter 2, Automating Machine Learning Model Development Using SageMaker Autopilot*.

2. Once the **SageMaker Domain** screen is open, make a note of the name of the **Execution** role in the **Domain** section. We will use the **Amazon Resource Name (ARN)** of the execution role to assign it the necessary permissions.

3. Now, open the **Identity and Access Management (IAM)** console (`https://console.aws.amazon.com/iam/home`) in a new browser tab.

4. On the left-hand navigation panel, click on **Roles**, under the **Access management** section, to open the **Roles** dashboard.

5. Find the execute role that you made a note of in *step 2* and click on it. The role name should start with **AmazonSageMaker-ExecutionRole-XXX**.

6. In the role's **Summary** dashboard, click on the **Trust relationships** tab and then the **Edit trust relationship** button.

7. Delete the existing **Policy Document** name and paste the following policy into the window. This will provide the execution role with trust access to both the SageMaker service and the CodeBuild service:

```
{
    "Version": "2012-10-17",
    "Statement": [
        {
            "Effect": "Allow",
            "Principal": {
                "Service": [
                    "codebuild.amazonaws.com",
                    "sagemaker.amazonaws.com"
                ]
            },
            "Action": "sts:AssumeRole"
        }
    ]
}
```

8. Click on the **Update Trust Policy** button.

Now that we have the necessary permissions to access the CodeBuild service, we can now use the Studio UI to prepare the custom SageMaker container.

Building a deep learning container

In order to build a custom container for AutoGluon, we need to provide detailed build instructions. For containers, these build instructions are included in a file called a Dockerfile.

> **Note**
>
> If you are unfamiliar with how to build Docker containers or how to construct a Dockerfile, you can refer to the Dockerfile reference documentation (https://docs.docker.com/engine/reference/builder/).

However, instead of building a `Dockerfile` from scratch, AWS provides pre-built container images and `Dockerfile` references for SageMaker, called Deep Learning Containers, or DL Containers (`https://aws.amazon.com/machine-learning/containers/`). These DL container images are engineered by AWS to support multiple deep learning frameworks (TensorFlow, PyTorch, and Apache MXNet) and are optimized for running ML use cases on the AWS cloud. Using these container images means you don't have to worry about configuring all the necessary Python dependencies and versions that these frameworks normally require. Alongside these deep learning frameworks, AWS also provides a pre-packaged DL container for AutoGluon (`https://github.com/aws/deep-learning-containers/blob/master/available_images.md#autogluon-inference-containers`).

The following steps will walk you through how to use the pre-built AutoGluon container and customize it for our requirements:

1. Using the **Amazon SageMaker** management console, click the **Open SageMaker Studio** button.

2. Click the **Open Studio** link to launch the Studio UI.

3. Within the Studio UI, click on the folder icon in the left sidebar.

4. Right-click in the folder navigation panel and click **New Folder**.

5. Name the new folder **Tabular** and double-click to open the folder.

6. In the menu bar, click **File | New |Notebook** and, when prompted, select the **Python 3 (Data Science)** kernel from the dropdown. Click the **Select** button to launch the **KernelGateway**. After a couple of minutes, the kernel will be ready.

> **Tip**
> You can create a new Jupyter notebook or use the example notebook from the companion GitHub repository (`https://github.com/PacktPublishing/Automated-Machine-Learning-on-AWS/blob/main/Chapter03/Tabular/AutoGluon%20Tabular%20Example.ipynb`).

7. In the first code cell, we will install the `sagemaker-studio-image-build` utility. Using the following code, we call the Python executable to run the Python package manager and install the utility:

```
%%capture
import sys
import warnings
```

```
warnings.filterwarnings('ignore')
%matplotlib inline
!{sys.executable} -m pip install -U pip sagemaker-studio-
image-build
```

8. Next, we open a new code cell to build the AutoGluon training and test script. This script is, in essence, the runtime that will be executed with the custom SageMaker container to generate and test the various AutoGluon tabular models to determine the best-fitting model. Using the following code, we won't execute this script inside the Jupyter notebook code cell, but rather use the %%writefile Jupyter magic command to create a script file called train.py:

```
%%writefile train.py
import os
import json
import boto3
import json
import warnings
import numpy as np
import pandas as pd
from autogluon.tabular import TabularDataset,
TabularPredictor

warnings.filterwarnings("ignore",
category=DeprecationWarning)
prefix = "/opt/ml"
input_path = os.path.join(prefix, "input/data")
output_path = os.path.join(prefix, "output")
model_path = os.path.join(prefix, "model")
param_path = os.path.join(prefix, 'input/config/
hyperparameters.json')
```

> **Tip**
> If you are unfamiliar with the Jupyter built-in magic commands, such as %%writefile, you can refer to the Jupyter documentation website (https://ipython.readthedocs.io/en/stable/interactive/magics.html#built-in-magic-commands) to learn more about these commands and how they can be used.

9. Using the following code, while still in the current code cell, we now define the `train()` function. This function takes in the various parameters, such as the prediction target label, and fits `TabularPredictor()` for this target, against `training_dataset`. Once the predictor has automatically determined the best models, we save the results as a `Fit_Summary.txt` file:

```python
def train(params):
    label = params["label"]
    channel_name = "training"
    training_path = os.path.join(input_path, channel_
name)
    training_dataset = TabularDataset(os.path.
join(training_path, "training.csv"))
    predictor = TabularPredictor(label=label, path=model_
path).fit(training_dataset)
    with open(os.path.join(model_path, "Fit_Summary.
txt"), "w") as f:
        print(predictor.fit_summary(), file=f)
    return predictor
```

10. In the following code, while still in the current code cell, we define the `test()` function. This function once again takes the target label, as well as the trained predictor, and evaluates the generated values on `testing_data`. The evaluation results are saved as a `Model_evaluation.txt` file. The `test()` function also generates a leaderboard of the best model and saves this list as a `Leaderboard. csv` file:

```python
def test(params, predictor):
    label = params["label"]
    channel_name = "testing"
    testing_path = os.path.join(input_path, channel_name)
    testing_dataset = TabularDataset(os.path.
join(testing_path, "testing.csv"))
    ground_truth = testing_dataset[label]
    testing_data = testing_dataset.drop(columns=label)
    predictions = predictor.predict(testing_data)
    with open(os.path.join(model_path, "Model_Evaluation.
txt"), "w") as f:
        print(
            json.dumps(
```

```
            predictor.evaluate_predictions(
                y_true=ground_truth,
                y_pred=predictions,
                auxiliary_metrics=True
            ),
            indent=4
        ),
        file=f
    )
    leaderboard = predictor.leaderboard(testing_dataset,
silent=True)
    leaderboard.to_csv(os.path.join(model_path,
"Leaderboard.csv"))
```

11. In the following code, and still within the current code cell, we define the main
program routine that loads the execution parameters, calls the `train()` function
to train the various predictors on the training data, and then calls the `test()`
function to evaluate the predictor performance on the test dataset:

```
if __name__ == "__main__":
    print("Loading Parameters\n")
    with open(param_path) as f:
        params = json.load(f)
    print("Training Models\n")
    predictor = train(params)
    print("Testing Models\n")
    test(params, predictor)
    print("AutoGluon Job Complete")
```

12. Next, we create a new code cell and apply the same technique, using the
`%%writefile` magic command, to create the custom container build instructions
or a `Dockerfile`. The following code contains the instructions that the Docker
daemon will use to build the container:

```
%%writefile Dockerfile
ARG REGION
FROM 763104351884.dkr.ecr.${REGION}.amazonaws.com/
autogluon-training:0.3.1-cpu-py37-ubuntu18.04
RUN pip install -U pip
```

```
RUN pip install bokeh==2.0.1
RUN mkdir -p /opt/program
RUN mkdir -p /opt/ml
COPY train.py /opt/program
WORKDIR /opt/program
ENTRYPOINT ["python", "train.py"]
```

13. Now, we can use the build CLI to create the customer container:

```
import boto3
import sagemaker
aws_region = sagemaker.Session().boto_session.region_name
!sm-docker build --build-arg REGION={aws_region} .
```

The container should take about 10 minutes to build, with the logs from the CodeBuild execution redirected to, and displayed in, the Jupyter notebook.

> **Note**
>
> Make sure to capture the **Image Uniform Resource Identifier (URI)** container from the code cell output, as we will be using this later. The code cell output should resemble this: `Image URI: 123456789012.dkr.ecr.us-west-2.amazonaws.com/sagemaker-studio-d-abcdefghij kl:default-1234567890123`.

You may be wondering exactly what the previous code cells accomplished. Firstly, let's walk through the `train.py` file. In this file, we've created two main Python functions, `train()` and `test()`:

- The `train()` function takes a training dataset called `training.csv` and creates a default AutoGluon Tabular predictor called `predictor`. The default predictor produces several different types of ML models that predict the target label by training on the other columns of the dataset. This process is similar to Autopilot's *Auto* setting, which was used in the example from *Chapter 2, Automating Machine Learning Model Development Using SageMaker Autopilot*. We will see later, when we execute the AutoGluon experiment, just exactly what these default models are and how well they perform on this training data.

- After the training process has been completed, the `train()` function returns these models as an AutoGluon `TabularPredictor` object.

The `test()` function takes the trained models as an input and then evaluates the various AutoGluon Tabular models, produced by the `train()` function, and evaluates them on a test dataset, called `testing.csv`. As a result of this process, the `test()` function stores the overall evaluation results for each model as well as a summary of the best model's score. As we will see later, these assets are eventually stored in S3 for review.

Now, let's review the `Dockerfile`. As already mentioned, the `Dockerfile` contains the instructions to build the Docker container that will execute the `train.py` script. The first build command to be executed is a command to download the AutoGluon DL container. The Docker daemon pulls this container from a public AWS ECR repository, called

`763104351884.dkr.ecr.${REGION}.amazonaws.com/autogluon-training:0.3.1-cpu-py37-ubuntu18.04`, where `${REGION}` is a build argument specifying the AWS region you are currently using.

> **Tip**
>
> For a list of the public ECR repositories containing the latest DLC images, refer to the project's GitHub repository (`https://github.com/aws/deep-learning-containers/blob/master/available_images.md`).

The Docker daemon then installs the necessary Python packages for AutoGluon, sets up the code path required by SageMaker, and copies the training script into the container.

> **Note**
>
> To install AutoGluon in the container, we have used the default installation requirements from the AutoGluon documentation (`https://auto.gluon.ai/stable/index.html#installation`).

In the last part of the Dockerfile, we specify the container's `ENTRYPOINT`, thus instructing the container to execute the training script when it starts.

Lastly, we executed the `sm-docker build` command, specifying the current AWS region as the build argument and the location of the Dockerfile in the current directory. Since we do not supply any further parameters, `sm-docker` assumes the default settings.

> **Tip**
>
> To see some of the additional settings that can be used instead of the defaults, refer to the utility documentation website (`https://github.com/aws-samples/sagemaker-studio-image-build-cli`).

Using the default settings, the CLI automatically calls AWS CodeBuild in the background. CodeBuild, in turn, executes the following tasks:

- It creates an ECR repository, named after the Studio Domain ID, for example, `sagemaker-studio-d-abcdefghijkl`.

- It builds the container image, with a unique image tag, for example, `default-1234567890123`.

- It uploads the image to the newly created repository.

> **Tip**
>
> To view the build process and configuration settings that the CodeBuild service executes, you can view and manage the process in the CodeBuild console (`https://console.aws.amazon.com/codesuite/codebuild/home`).

Now that the build process has been completed, we have, in essence, brought out our own AutoGluon container to SageMaker. In the next section, we will use this container to conduct an AutoML experiment for the *ACME Fishing Logistics* use case.

Creating the AutoML experiment with AutoGluon

In the same way that we created an AutoML experiment in *Chapter 2, Automating Machine Learning Model Development Using SageMaker Autopilot*, using the SageMaker SDK, we will reproduce a similar experiment by working with AutoGluon. The following steps will walk you through creating the experiment within the existing Jupyter notebook:

1. Firstly, we need to download the Abalone dataset once again. Using the following Python code, download the dataset from the UCI repository, add the relevant column names, and split the data into two separate CSV files, `training.csv` and `testing.csv`. The training file comprises 90% of the data, while the testing file covers the remaining 10%. As already highlighted, these two datasets will be used by the `train()` and `test()` functions within our container. In a new code cell, execute the following example code:

```
import numpy as np
import pandas as pd
from sklearn.model_selection import train_test_split
column_names = ["sex", "length", "diameter", "height",
"whole_weight", "shucked_weight", "viscera_weight",
"shell_weight", "rings"]
```

```
abalone_data = pd.read_csv("http://archive.ics.uci.edu/
ml/machine-learning-databases/abalone/abalone.data",
names=column_names)
training_data, testing_data = train_test_split(abalone_
data, test_size=0.1)
training_data.to_csv("training.csv")
testing_data.to_csv("testing.csv")
```

2. Now that we have our datasets, we can configure the various parameters for the experiment. Using the following code, we can define important parameters, such as the name of the experiment (`job_name`), the specific version to trace the experiment (`job_version`), the SageMaker default S3 bucket to store the datasets, the output artifacts (`bucket`), and the container URI (`image_uri`):

```
import sagemaker
import datetime
image_uri = "<Enter the Image URI from the sm-docker
output>"
role = sagemaker.get_execution_role()
session = sagemaker.session.Session()
bucket = session.default_bucket()
job_version = datetime.datetime.now().strftime("%Y-%m-%d-
%H-%M-%S-%f")[:-3]
job_name = f"abalone-autogluon-{job_version}"
```

> **Note**
>
> For the `image_uri` parameter, enter the URI from the output of the sm-docker code cell we executed earlier in *step 10* of the previous section.

3. Using a new code cell, we use these parameters and the SageMaker SDK to create a SageMaker estimator. An estimator is a high-level interface for a SageMaker training job. The following code uses the generic `sagemaker.estimator.Estimator()` class, allowing us to create a training job using our custom AutoGluon container. As you can see, we also supply additional hyperparameters where we specify the type of compute instance (`ml.m5.xlarge`) to use to execute the training job as well as the parameters to be supplied to the `train.py` script, such as the target label in our dataset (`rings`):

```
from sagemaker.estimator import Estimator
autogluon = Estimator(
```

```
        image_uri=image_uri,
        role=role,
        output_path=f"s3://{bucket}/{job_name}",
        base_job_name=job_name,
        instance_count=1,
        instance_type="ml.m5.xlarge",
        hyperparameters={
            "label": "rings",
            "bucket": bucket,
            "training_job": job_name
        },
        volume_size=20
)
```

> **Tip**
>
> For more information on the generic Estimator() class, refer to
> the SageMaker SDK documentation (https://sagemaker.
> readthedocs.io/en/stable/api/training/estimators.
> html#sagemaker.estimator.Estimator).

4. Now that the estimator has been defined, we can use the fit() method to call
    SageMaker and have it execute the training job, using our custom AutoGluon
    container. As you can see in the following code, we tell SageMaker where to get
    the training and test data by uploading these datasets to S3, using the upload_
    data() method:

```
autogluon.fit(
    inputs={
        "training": session.upload_data(
            "training.csv",
            bucket=bucket,
            key_prefix=f"{job_name}/input"
        ),
        "testing": session.upload_data(
            "testing.csv",
            bucket=bucket,
            key_prefix=f"{job_name}/input"
        )
```

```
        }
    )
```

After calling the `fit()` method, SageMaker will instantiate an `ml.m5.xlarge` instance with **4vCPUs** and **16 GB** of RAM to execute our custom AutoGluon container. The output from the inside of the container's runtime environment is redirected to, and displayed in, the Jupyter notebook. You can review each line of the output to see what is going on. Alternatively, *Figure 3.1* provides a high-level overview of the output from the SageMaker training job:

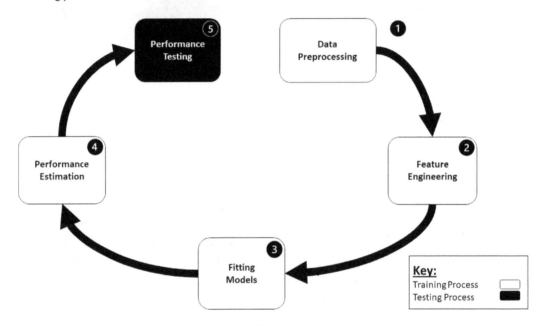

Figure 3.1 – AutoGluon process overview

As you can see from *Figure 3.1*, five specific steps are executed by the `train()` function and a single step is performed by the `test()` function. Let's examine each step and correlate it with the output:

1. The first step that AutoGluon does is to preprocess the data. Here, AutoGluon analyzes the data to try and determine the type of ML problem. For example, AutoGluon may determine that the ML problem is multiclass because the target label's data type is an integer and there are very few unique values observed. Once AutoGluon has determined the ML problem type, it further performs preprocessing of the input data for the specific model.

> **Tip**
>
> If you already know the type of ML you are trying to solve, you can specifically add it as an argument, called `problem_type`, to AutoGluon's `fit()` method.

2. The second step involves AutoGluon using several pre-built data generators to clean up data and engineer new features. For example, AutoGluon uses `FillNaFeatureGenerator` to automatically determine the type of values to replace any missing values present in the dataset, plus `CategoryFeatureGenerator` to encode categorical features. The final part of this step involves splitting the processed dataset into separate training and testing sets.

3. In the third step, AutoGluon trains its pre-built ML models for tabular data against the training dataset. For a list of the 10 specific models and the 3 ensemble models that AutoGluon Tabular fits on the training set, refer to the AutoGluon documentation (`https://auto.gluon.ai/stable/api/autogluon.tabular.models.html#module-autogluon.tabular.models`).

4. The final step of the `train()` process evaluates the trained models against the validation dataset to determine the model's overall validation score and see how each model generalizes to accurately predict the target label. The default metric for evaluation is **accuracy**. Each model is serialized and stored as a Python object using the `pickle` library.

> **Tip**
>
> To change the default evaluation metric, you can specify it as an argument, called `eval_metric`, to AutoGluon's `fit()` method.

5. To complete the process, the `test()` function is then executed to provide a final evaluation on unseen data, using the `testing.csv` dataset. This final step provides us with the overall performance of each of the trained models to generate the best model. The final results of the evaluation are captured as an output artifact along with the *pickled* models.

Once SageMaker has executed the `train.py` script, the models and evaluation summary artifacts are compressed and uploaded to S3. SageMaker displays how long the training job took to execute and, hence, the total amount of **Billable seconds**.

Now that we have created and implemented the AutoGluon Tabular experiment on the Abalone dataset, we can evaluate the models generated to determine which can be used in production. The next section will show how this is done.

Evaluating the experiment results

As was highlighted in the previous section, the model evaluation results and *pickled* models are captured as an output artifact, called `model.tar.gz`, and uploaded to S3. Using the existing Jupyter notebook, let's take a look at these artifacts to assess the results of the AutoML experiment and determine which model best suits the production use case:

1. The following example code uses the SageMaker SDK's `S3Downloader` class to download and extract the AutoGluon estimator's output artifact to a folder called `extract`, using the `model_data` property:

    ```
    !mkdir extract sagemaker.s3.S3Downloader.
    download(autogluon.model_data, "./")
    !tar xfz ./model.tar.gz -C extract
    ```

2. You can look through the extracted contents, in the `extract` folder, to see the various evaluation reports, and double-click on the `models` folder to see the *pickled* model artifacts. *Figure 3.2* shows the extracted artifact files:

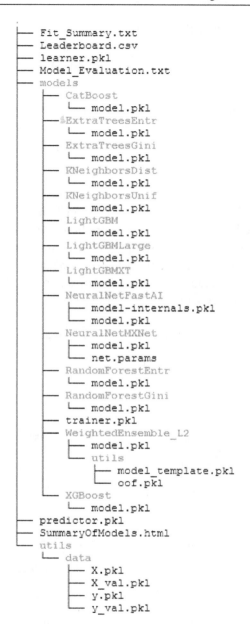

```
├── Fit_Summary.txt
├── Leaderboard.csv
├── learner.pkl
├── Model_Evaluation.txt
├── models
│   ├── CatBoost
│   │   └── model.pkl
│   ├── ExtraTreesEntr
│   │   └── model.pkl
│   ├── ExtraTreesGini
│   │   └── model.pkl
│   ├── KNeighborsDist
│   │   └── model.pkl
│   ├── KNeighborsUnif
│   │   └── model.pkl
│   ├── LightGBM
│   │   └── model.pkl
│   ├── LightGBMLarge
│   │   └── model.pkl
│   ├── LightGBMXT
│   │   └── model.pkl
│   ├── NeuralNetFastAI
│   │   ├── model-internals.pkl
│   │   └── model.pkl
│   ├── NeuralNetMXNet
│   │   ├── model.pkl
│   │   └── net.params
│   ├── RandomForestEntr
│   │   └── model.pkl
│   ├── RandomForestGini
│   │   └── model.pkl
│   ├── trainer.pkl
│   ├── WeightedEnsemble_L2
│   │   ├── model.pkl
│   │   └── utils
│   │       ├── model_template.pkl
│   │       └── oof.pkl
│   └── XGBoost
│       └── model.pkl
├── predictor.pkl
├── SummaryOfModels.html
└── utils
    └── data
        ├── X.pkl
        ├── X_val.pkl
        ├── y.pkl
        └── y_val.pkl
```

Figure 3.2 – Extracted artifact folder structure

3. The first file we will look at is the `Leaderboard.csv` file and see the overall performance evaluation of each of the trained models. The following code opens the model leaderboard as a pandas DataFrame and sorts the models in descending order:

```
df = pd.read_csv("./extract/Leaderboard.csv")
df = df.filter(["model", "score_test", "score_val"]).
sort_values(by="score_val", ascending=False).reset_
index().drop(columns="index")
df
```

4. You can now review the models that the AutoGluon Tabular predictor trained. The best model, based on the accuracy of its predictions on the test dataset, is displayed first. *Figure 3.3* shows an example of the leaderboard table and, as you can see, `WeightedEnsemble_L2` (https://auto.gluon.ai/stable/api/autogluon.tabular.models.html#weightedensemblemodel) provided the *best* validation accuracy score (`score_val`).

	model	score_test	score_val
0	WeightedEnsemble_L2	0.229665	0.307229
1	XGBoost	0.229665	0.303213
2	LightGBM	0.248804	0.289157
3	NeuralNetMXNet	0.296651	0.275100
4	NeuralNetFastAI	0.284689	0.273092
5	LightGBMLarge	0.253589	0.271084
6	LightGBMXT	0.267943	0.267068
7	RandomForestEntr	0.265550	0.263052
8	CatBoost	0.263158	0.257028
9	RandomForestGini	0.251196	0.255020
10	ExtraTreesGini	0.263158	0.251004
11	ExtraTreesEntr	0.260766	0.240964
12	KNeighborsUnif	0.217703	0.220884
13	KNeighborsDist	0.200957	0.218876

Figure 3.3 – Example model leaderboard

> **Note**
>
> AutoGluon generates evaluation metric scores in a *higher is better* form.
> Therefore, the higher the evaluation score, the better the model.

5. Just like the Autopilot example in *Chapter 2, Automating Machine Learning Model Development Using SageMaker Autopilot*, we can visually compare the models in the leaderboard with code. However, AutoGluon Tabular automatically constructs a model comparison plot, as an output artifact, called `SummaryOfModels.html`. The following example code will display the plot in the Jupyter notebook:

```
import IPython
IPython.display.HTML(filename="./extract/SummaryOfModels.
html")
```

> **Note**
>
> If the `SummaryofModels.html` file does not display immediately when running the code in *step 5*, rerun the code cell again.

6. *Figure 3.4* shows an example of the displayed `SummaryOfModels.html` file. Interact with the plot by mousing over the generated scatterplot and viewing the metadata for each of the models.

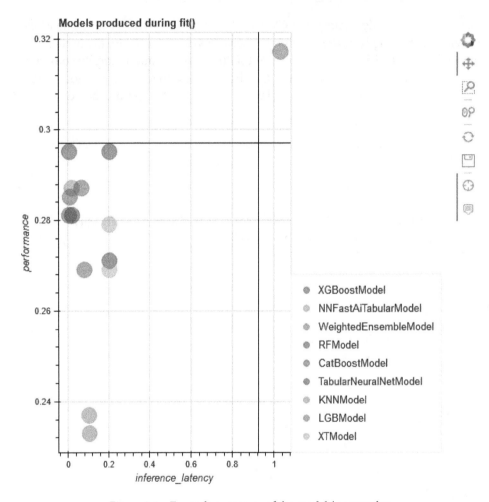

Figure 3.4 – Example summary of the models' scatterplot

Once again, we have used an AutoML methodology, this time using the AutoGluon Tabular predictor, to create a feasible production-grade ML model for our use case. As with the *typical* ML process, an ML practitioner can provide the *best* model to the application teams for testing and integration into the *Age Calculator* application.

One thing you may be wondering is why we didn't simply execute the AutoGluon training and evaluation process inside the existing Jupyter notebook. *Why did we create a custom container and run the entire process as a SageMaker training job?*

The answer to this question is basically cost. To elaborate, in the next example, we will review a use case that requires **Graphical Processing Units** (**GPUs**) to train and evaluate intricate computer vision models. This use case will highlight the benefits of offloading resource-intensive model training jobs to SageMaker, using the AutoGluon `ImagePredictor`.

Using AutoGluon for image data

Up to this point, we have been exploring AutoML methodologies on an **Artificial Neural Network** (**ANN**) algorithm. However, many use cases might require more complicated algorithms, such as **Convolutional Neural Networks** (**CNNs**) for image classification and image recognition, or **Long-Short-Term Memory** (**LSTM**) networks, for speech recognition and text data. Due to the complexity of these algorithms, many ML practitioners may have to leverage multiple machines for distributed training and potentially multiple GPUs to handle the multi-dimensional matrix calculations. In this section, we are going to segue from the *Age Calculator* use case to explore how AutoGluon can be used to apply an AutoML methodology to an image classification use case.

> **Note**
>
> Since we will be utilizing GPU-based AWS instances, running this example will exceed the usage limits of the AWS Free Tier and, therefore, incur additional costs.

Prerequisites

As was the case with the previous example, to leverage GPUs for the image classification task, we will need to build a custom container. Once again, AWS provides a DL container for Apache MXNet with GPU support. So, all we need to do is build the appropriate AutoGluon runtime into the pre-built container.

> **Tip**
>
> You can create a new Jupyter notebook or use the example notebook from the companion GitHub repository (`https://github.com/PacktPublishing/Automated-Machine-Learning-on-AWS/blob/main/Chapter03/Image/AutoGluon%20Image%20Example.ipynb`).

The following steps will walk you through how to do this:

1. Within the same Studio environment from the previous example, click the folder icon in the left sidebar.

2. Right-click in the folder navigation panel and click **New Folder**.

3. Name the new folder `Image` and double-click to open the folder.

4. In the menu bar, click **File | New | Notebook** and, when prompted, select the **Python 3 (Data Science)** kernel from the dropdown. Click the **Select** button to launch the **KernelGateway**. After a couple of minutes, the kernel will be ready.

5. In the first code cell, we will once again install the `sagemaker-studio-image-build` utility by executing the following code:

```
%%capture
import sys
import warnings
warnings.filterwarnings('ignore')
%matplotlib inline
!{sys.executable} -m pip install -U pip sagemaker-studio-image-build
```

6. Next, we will build the AutoGluon training and test script. This script is, in essence, the runtime that will be executed with the custom SageMaker container to generate and test the various AutoGluon `ImagePredictor` models to determine the best-fitting model. Since we are capturing the contents of the code cell to a file, we use the `%%writefile` Jupyter magic command to create a script file called `train.py`:

```
%%writefile train.py
import os
import json
import boto3
import json
import warnings
import numpy as np
import pandas as pd
from autogluon.vision import ImagePredictor

warnings.filterwarnings("ignore",
category=DeprecationWarning)
```

```
prefix = "/opt/ml"
input_path = os.path.join(prefix, "input/data")
output_path = os.path.join(prefix, "output")
model_path = os.path.join(prefix, "model")
param_path = os.path.join(prefix, "input/config/
hyperparameters.json")
```

7. Within the same code cell, we define a `train()` function to capture the input parameters and fit an `ImagePredictor()` to `training_data`. We also capture a summary of the training results in a file called `FitSummary.csv` and save the trained model:

```
def train(params):
    time_limit = int(params["time_limit"])
    presets = "".join([str(i) for i in
list(params["presets"])])
    channel_name = "training"
    training_path = os.path.join(input_path, channel_
name)
    training_dataset = ImagePredictor.Dataset.from_
folder(training_path)
    predictor = ImagePredictor().fit(training_dataset,
time_limit=time_limit, presets=presets)
    with open(os.path.join(model_path, "FitSummary.
json"), "w") as f:
        json.dump(predictor.fit_summary(), f)
    predictor.save(os.path.join(model_path,
"ImagePredictor.Autogluon"))
    return "AutoGluon Job Complete"
```

8. Lastly, within the same code cell, we define the main program to load the input parameters and execute the model's training by calling the `train()` function and capturing the results:

```python
if __name__ == "__main__":
    print("Loading Parameters\n")
    with open(param_path) as f:
        params = json.load(f)
    print("Training Models\n")
    result = train(params)
    print(result)
```

9. As with the tabular example, we provide the build instructions for the custom container by creating a `Dockerfile`:

```
%%writefile Dockerfile
ARG REGION
FROM 763104351884.dkr.ecr.${REGION}.amazonaws.com/
autogluon-training:0.3.1-gpu-py37-cu102-ubuntu18.04
RUN mkdir -p /opt/program
RUN mkdir -p /opt/ml
COPY train.py /opt/program
WORKDIR /opt/program
ENTRYPOINT ["python", "train.py"]
```

> **Note**
>
> Once again, we are using the `autogluon-training` container, provided by AWS (`https://github.com/aws/deep-learning-containers/blob/master/available_images.md`), but this time, we will be using the GPU version of the image, denoted by the `0.3.1-gpu-py37-cu102-ubuntu18.04` image tag. Using the DL containers means that we don't have to manually build and configure the GPU environment, CUDA libraries (`https://blogs.nvidia.com/blog/2012/09/10/what-is-cuda-2/`), and runtime since AWS has done this for us.

10. Now, we can use the build CLI to create the customer container:

```python
import boto3
import sagemaker
```

```
aws_region = sagemaker.Session().boto_session.region_name
!sm-docker build --build-arg REGION={aws_region} .
```

The container should take about 10 minutes to build, with the logs from the CodeBuild execution redirected to, and displayed in, the Jupyter notebook.

Note

As was the case with the previous example, make sure to capture the **Image Uniform Resource Identifier (URI)** container from the code cell output as we will be using this later. The code cell output should resemble this: `Image URI: 123456789012.dkr.ecr.us-west-2.amazonaws.com/sagemaker-studio-d-abcdefghijkl:default-1234567890123`.

As you can see, the procedures executed in the example code closely resemble the procedures we ran for the Tabular example, except for the `train.py` script. Here, we make use of AutoGluon's `ImagePredictor` class, instead of the `TabularPredictor` class, whereby the `train()` function in this example takes a list of presets and fits multiple pre-trained and highly accurate CNN models on the image dataset provided. The `fit()` method automatically tries to improve the classification accuracy of the pre-trained models by employing additional hyperparameter optimization techniques, with the result being an optimized model and a set of model optimization parameters for reproducibility.

Unlike the Tabular example, we haven't made use of a `test()` function, since the image predictor automatically splits the image dataset into training and validation datasets, using a *90%/10%* split ratio.

Let's see this in action by creating an experiment in the next section.

Creating an image prediction experiment

For this experiment, we will be using the *Rock Paper Scissors* dataset that has kindly been provided by *Laurence Moroney* (`https://laurencemoroney.com/datasets.html`).

Note

This dataset is licensed under a Creative Commons 2.0 Attribution 2.0 Unported License (`https://creativecommons.org/licenses/by/2.0/`).

This dataset includes **Computer-Generated Imagery (CGI)** of different hand gestures, indicating either a Rock, Paper, or Scissors pose. *Figure 3.5* shows an example of each of these poses:

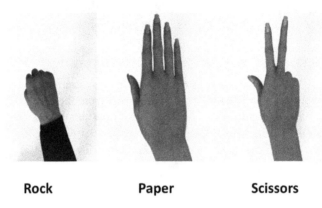

Rock   **Paper**   **Scissors**

Figure 3.5 – Examples of the Rock Paper Scissors dataset

In the same way that we created a tabular experiment using the SageMaker SDK, we will reproduce a similar experiment by working through the following steps in the existing Jupyter notebook:

1. The first step is to download the training image data from Laurence Moroney's website (https://storage.googleapis.com/laurencemoroney-blog.appspot.com/rps.zip). Since the dataset is provided in a compressed ZIP file, we will also need to extract it locally. The following sample code shows how this is accomplished:

```python
import io
import urllib
import zipfile

dataset_url = "https://storage.googleapis.com/
laurencemoroney-blog.appspot.com/rps.zip"
with urllib.request.urlopen(dataset_url) as rps_zipfile:
    with zipfile.ZipFile(io.BytesIO(rps_zipfile.read()))
as z:
        z.extractall("data")
```

2. Once the dataset has been downloaded, you should see a data folder containing the various sub-directories for each type or classification of image. For example, you will see a sub-directory called rock, which contains the training images depicting the pose for rock. AutoGluon will automatically use these sub-directories as the target label with which to classify the images. Next, we configure the various parameters for the experiment. Using the following code, we can define important parameters, such as the name of the experiment (job_name), the specific version to trace the experiment (job_version), the SageMaker default S3 bucket to store the datasets and the output artifacts (bucket), and the container URI (image_uri):

```
import sagemaker
import datetime


image_uri = "<Enter the Image URI from the sm-docker
output>"
role = sagemaker.get_execution_role()
session = sagemaker.session.Session()
bucket = session.default_bucket()
job_version = datetime.datetime.now().strftime('%Y-%m-%d-
%H-%M-%S-%f')[:-3]
job_name = f"autogluon-image-{job_version}"
```

> **Note**
>
> Make sure to enter the URI from the output of the sm-docker code cell we executed in *step 8* of the previous section.

3. Now that we have the various experiment parameters configured, we can create the AutoGluon estimator. The following example code applies a process similar to the tabular example, except for some of the hyperparameters:

```
from sagemaker.estimator import Estimator


autogluon = Estimator(
    image_uri=image_uri,
    role=role,
    output_path=f"s3://{bucket}/{job_name}",
    base_job_name=job_name,
    instance_count=1,
```

```
    instance_type="ml.p2.xlarge",
    hyperparameters={
        "presets": "medium_quality_faster_train",
        "time_limit": "600",
        "bucket": bucket,
        "training_job": job_name
    },
    volume_size=50
)
```

4. The final step in executing the experiment is to call SageMaker and have it execute the training job. Just as with the Tabular example, the following code executes the `fit()` method on the estimator and tells SageMaker where to get the image data by uploading the dataset to S3:

```
autogluon.fit(
    inputs={
        "training": session.upload_data(
            "data/rps",
            bucket=bucket,
            key_prefix=f"{job_name}/input"
        )
    }
)
```

Once the `fit()` method has been called, SageMaker will provision a GPU-based instance (`ml.p2.xlarge`), initialize the GPU-based container image, and execute the `train()` function. As part of this process, the AutoGluon `ImagePredictor` will determine the number of separate image classes, based on the sub-directories in the dataset, and start downloading various pre-trained CNN models to execute hyperparameter optimization tasks. The specific pre-trained models and hyperparameters are governed by the **presets** we've defined. For example, when creating the estimator, we specified `medium_quality_faster_train` as one of the presets. This preset will only use the *resnet50* pre-trained model to provide medium predictive accuracy as well as very fast inference and training times.

> **Note**
>
> We chose to use the `medium_quality_faster_train` preset and set a time limit of 10 minutes (`600` seconds) to reduce the amount of AWS usage costs associated with running the experiment. AutoGluon provides a number of alternative presets that will provide a better-quality model but incur additional AWS usage costs. You can learn more about the additional preset configurations that are available by referencing the `ImagePredictor` documentation (`https://auto.gluon.ai/ dev/api/autogluon.task.html#autogluon.vision. ImagePredictor.fit`).

When the training job is complete, the next step is to evaluate the result.

Evaluating the experiment results

As we saw in the tabular example, SageMaker will store the resulting model artifact in S3. Continuing in the Jupyter notebook, the following steps will walk you through how to evaluate the AutoML experiment:

1. Download and extract the `model.tar.gz` artifact to a folder called `extract` within the Studio environment by running the following code:

    ```
    !mkdir extract
    sagemaker.s3.S3Downloader.download(autogluon.model_data,
    "./")
    !tar xfz ./model.tar.gz -C extract
    ```

2. The model artifact contains two files, `ImagePredictor.Autogluon` and `FitSummary.json`. We can explore the model training summary by running the following code and viewing the `FitSummary.json` file:

    ```
    import json
    with open("extract/FitSummary.json", "r") as f:
        fit_summary = json.load(f)
    print(json.dumps(fit_summary, indent=4))
    print(f"""Best Model Training Accuracy: {fit_
    summary["train_acc"]} \nBest Model Validation Accuracy:
    {fit_summary["valid_acc"]}""")
    ```

After executing the previous code cell, you should see an output similar to the following JSON snippet:

```json
...
    "best_config": {
        "model": "resnet50d",
        "lr": 0.01,
        "num_trials": 1,
        "epochs": 50,
        "batch_size": 64,
        "nthreads_per_trial": 128,
        "ngpus_per_trial": 8,
        "time_limits": 600,
        "search_strategy": "random",
        "dist_ip_addrs": null,
        "log_dir": "/opt/program/85cde890",
        "searcher": "random",
        "scheduler": "local",
        "early_stop_patience": 5,
        "early_stop_baseline": -Infinity,
        "early_stop_max_value": Infinity,
        "num_workers": 4,
        "gpus": [0],
        "seed": 206,
        "final_fit": false,
        "wall_clock_tick": 1640286670.9024706,
        "problem_type": "multiclass"
    },
...
```

As you can see from this JSON snippet, the `resnet50d` model provided the best configuration. Included in the JSON snippet is the best hyperparameter configuration for us to reproduce the model again without having to run another AutoML experiment. Additionally, if you review the last few lines of the JSON output, you will see the evaluation results of the best model. The following snippet shows an example of the model's accuracy metrics:

```
...
Best Model Training Accuracy: 0.8929924242424242
Best Model Validation Accuracy: 0.996031746031746
...
```

From this final snippet, you can see that the `resnet50d` model achieved an **89%** accuracy on the training dataset, and a **99%** accuracy on the validation dataset. So, depending on this use case, these metrics might qualify the model to be put into production, and therefore the `ImagePredictor.Autogluon` artifact, stored in the `extract` folder, can be provided to the application development teams.

So, by means of this example, we have accomplished two main goals:

- We have created an AutoML experiment to address a complex use case (Computer Vision), requiring a more complicated ML algorithm, such as a CNN model. Just as with the tabular example, we used the AutoGluon library to generate the best-fitting model for the image data.

- While the Studio UI provides the capability to run a GPU-based **KernelGateway** (https://docs.aws.amazon.com/sagemaker/latest/dg/notebooks-available-instance-types.html), we further alleviated unnecessary AWS costs to run the Jupyter notebook while exploring how to run the AutoGluon model and configure the various CUDA libraries for GPU management. Instead, we created a training *runtime artifact*, as a custom image from the pre-built DL container, and offloaded the AutoML processing to a SageMaker training job.

In later chapters, we will leverage the technique of building a runtime artifact to further streamline the ML automation process.

Summary

This chapter introduced you to an open source alternative for creating an AutoML process using the AutoGluon Python library. We also used AutoGluon's Tabular predictor to advance the *Age Calculator* use case and demonstrated how to find the best-suited model for the tabular dataset.

We further expanded on the AutoML methodology to address a complicated computer vision use case by finding the best-suited CNN model for the *Rock Paper Scissors* dataset. This was accomplished using AutoGluon's Image predictor and further optimized using SageMaker's GPU-based ML instances. This chapter also introduced the concept of a runtime process artifact, in the form of a container image.

In the next chapter, we will continue to expound on this concept and introduce how an ML runtime artifact can further streamline the ML process, especially when the artifact is used in conjunction with other AWS services.

Section 2: Automating the Machine Learning Process with Continuous Integration and Continuous Delivery (CI/CD)

This section will introduce you to the concepts of CI/CD, and how they can be applied to the ML process, by combining both DevOps and MLOps methodologies. We will showcase the various AWS services that can be used to build and execute a CI/CD pipeline for the ML process. This section will walk you through how to construct the CI/CD pipeline as a cloud-native application using the **Cloud Development Kit (CDK)**.

This section comprises the following chapters:

- *Chapter 4, Continuous Integration and Continuous Delivery (CI/CD) for Machine Learning*
- *Chapter 5, Continuous Deployment of a Production ML Model*

4
Continuous Integration and Continuous Delivery (CI/CD) for Machine Learning

While working through the code examples, in both *Chapter 2, Automating Machine Learning Model Development Using SageMaker Autopilot*, and *Chapter 3, Automating Complicated Model Development with AutoGluon*, for the *age calculator* use case, you would've noticed a common trend that highlighted a drawback in using either the Autopilot or AutoGluon methodologies – specifically, that there is a disconnect in both processes between creating a production-grade ML (**machine learning**) model and then actually deploying the model into production.

Whether an ML practitioner leverages the **CRISP-DM** methodology or an **AutoML** methodology, the scope of their responsibilities ends once they have produced an optimal ML model. After their task is complete, the ML practitioner simply hands the model over to the various teams responsible for deploying and managing the model in production. This handover creates a disconnect in the overall process and leads to further challenges when trying to automate the overall process. More importantly, this disconnect can often impact the overall delivery timeline and cause a delay in the successful completion of the overall project.

The primary goal of this chapter is to highlight one of the ways to bridge this apparent gap in model deployment and further automate the process, using a **Continuous Integration and Continuous Delivery (CI/CD)** methodology. I'm also going to introduce you to the concept of an agile, cross-functional team by showing you how an ML practitioner can better interface with the application development and operations teams, and by the end of the chapter, you will see how this methodology can consistently create production-grade ML models and deploy them. To accomplish this, we will focus on the following topics:

- Introducing the CI/CD methodology
- Automating ML with CI/CD
- Creating a CI/CD pipeline on AWS

Technical requirements

This chapter will use the following resources:

- A web browser (for the best experience, it is recommended that you use the Chrome or Firefox browser).
- Access to the AWS account that you used in *Chapter 3, Automating Complicated Model Development with AutoGluon.*
- Access to an **Integrated Development Environment** (**IDE**) if you choose not to use the AWS Cloud9 service.
- We will once again be working within the usage limits of the AWS Free Tier to avoid exceeding unnecessary costs.
- Source code example, access policy documents, and Jupyter notebooks are provided in the companion GitHub repository for this chapter (`https://github.com/ PacktPublishing/Automated-Machine-Learning-on-AWS/tree/ main/Chapter04`).

Introducing the CI/CD methodology

The CI/CD pattern has become a very popular methodology to automate the development and release of software into production. The main idea behind this practice is to make incremental, reliable, and frequent software code changes, and then deploy these changes automatically and seamlessly into production.

While this practice has been around for several years and employed by many DevOps engineers, the practice is starting to gain traction within the ML practitioner community, in the form of **MLOps** or **Machine Learning Operations**. However, before diving into how this methodology can be applied to ML, let's familiarize ourselves with the specific steps of the process, starting with CI.

Introducing the CI part of CI/CD

At a high level, the CI part of CI/CD comprises four key stages; *Figure 4.1* shows a high-level overview of what these stages are:

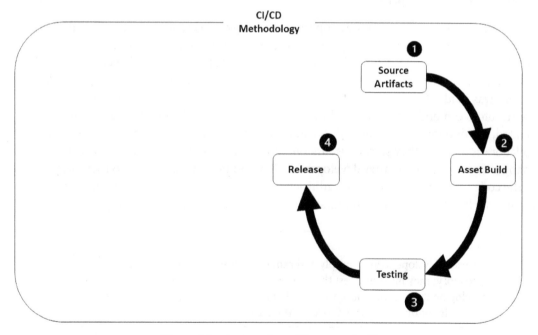

Figure 4.1 – An overview of the stages in CI

As you can see from *Figure 4.1*, the CI phase comprises the following key stages:

1. Source Artifacts
2. Asset Build
3. Integration Testing
4. Release

Let's review exactly what the steps of this process (or pipeline) entail.

Creating or updating source artifacts

The source artifacts stage doesn't really perform a specific task within the CI process, other than to start the entire process. In essence, this stage serves as a repository where developers store the source code or the pieces of software that comprise the production application. Adding new software artifacts (such as new features) or updating existing software artifacts (such as bug fixes) into this repository triggers the start of the entire pipeline.

For example, when application developers make code changes, add new features, or fix application bugs, they add these updates to the shared version control system (such as GitHub, Bitbucket, or AWS CodeCommit). These saved changes are called commits, with each commit having an associated description or message that explains why a particular change was made. These commits sum up the history of all the changes so that other contributors can understand what's been done to the code and why. Once a commit is created, the developer can open a **Pull Request** (**PR**). PRs are the nucleus of developer collaboration in that they start a discussion between team members over proposed changes and request that other developers review and pull the updates into their branch of the code repository. Once these additions are approved, they are then merged into the main branch of the repository to initiate or trigger a build of an updated application.

> **Note**
>
> Software developers don't simply add random features or updates to a repository; they must first test the new code in their local or cloud-based development environment to validate that the new code is functional. This process is commonly referred to as **unit testing**.

Building the pipeline assets

The next stage of the pipeline is where the various software artifacts (and their dependencies) are compiled or built. Essentially, these assets are, at a high level, the result of building the source code artifacts into an asset that is specific to the current release, or, in other words, assets that are specific to the current execution of the pipeline. For example, the build stage compiles C++ code into a release binary or the build stage can be used to build a Docker container image.

Testing the pipeline assets

Once the pipeline assets have been built, the next stage of the pipeline is to not only test that these assets are functional but also test that they **fit** into the overall architecture or application. It's at this stage that developers leverage testing scripts, automated testing, or even a testing architecture (called a **test** or QA environment) to perform **system testing**. The primary goal of this step is to verify that built assets will function correctly once they have been deployed into production. By testing the overall **system**, application developers can be assured that the overall integrity of the solution, in its entirety, is maintained once it's deployed into production.

Approving the release

Once the overall system or application integrity has been tested, the final part of the CI phase is to approve it for production. This stage of the process can either consist of a person (or team) approving the results of the test or, in the case of frequent code changes, be automated.

Once the release is approved, the CI phase of the pipeline is complete, and the release can then move onto the CD phase for production deployment. Let's review what the CD phase entails.

Introducing the CD part of CI/CD

The CD phase of the pipeline is just a continuation of the CI phase and is comprised of four individual stages that focus on the operational tasks of the production application. The four stages that constitute the CD phase of the CI/CD pipeline are as follows:

1. Asset Deployment
2. Operations
3. Monitoring
4. Operational Feedback

Figure 4.2 shows an overview of these four stages:

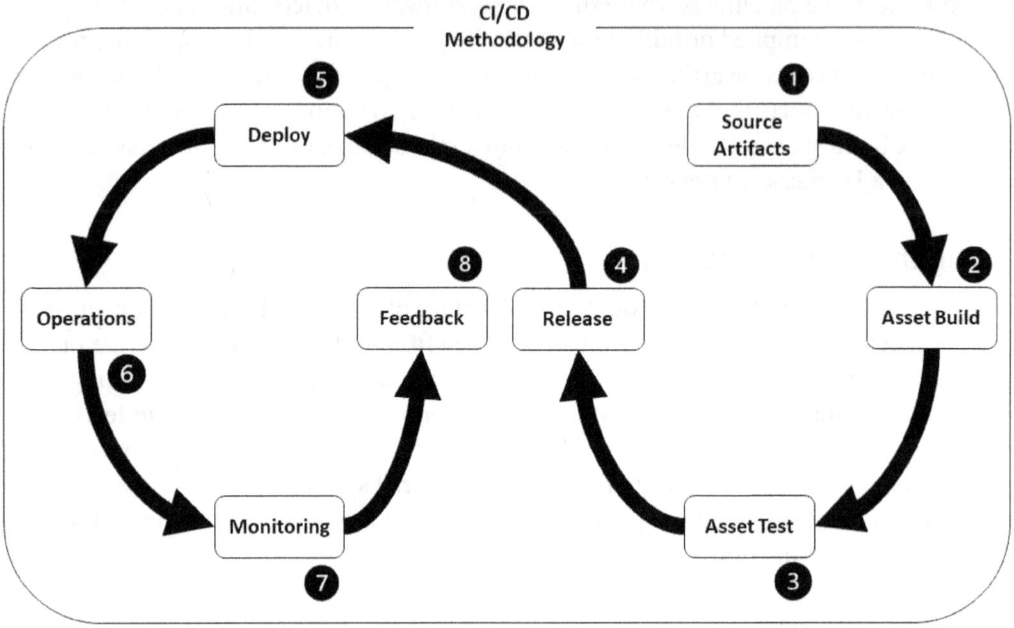

Figure 4.2 – An overview of the stages in CD

Let's look at what these stages entail.

Deploying the release into production

When deploying the built and tested software into production, there are two primary components of the process. The first component is a deployment process, while the second component is a deployment strategy.

For example, a deployment strategy may involve deploying a duplicate application into production and, over a period of time, redirecting new usage requests to the new release, while eventually phasing out the older release. This strategy is often called a **blue/ green deployment**.

On the other hand, a deployment process is an underlying mechanism of putting a new release into production. This process varies depending on the type of software or compiled asset being deployed. For example, if the deployed asset is a container image, the deployment process might involve downloading and running the container image by means of a container orchestration solution, such as Kubernetes.

Managing and monitoring the solution

To ensure that the solution functions the way it is supposed to, there are multiple tasks that are typically performed at the operations stage, which can also overlap with application monitoring tasks. So, typically, the operations and monitoring tasks are performed by the infrastructure or IT team at the same time. For example, these tasks might include updating underlying operating system patches, or ensuring that the architecture automatically scales to address an increase in usage by monitoring application performance.

Production feedback reporting

The feedback stage is also an extension of the management and monitoring tasks; however, it also involves parsing the various logs and reporting dashboards to isolate any failures, bugs or, issues from the production application. For example, this stage can involve looking for application errors from the applicable logs and generating a bug report. However, simply cataloging the bugs accomplishes nothing if the information is not communicated back to the application developers.

Therefore, the CI/CD process does not end at this stage. So, in the next section, we will look at how this feedback report is used to close the loop and ensure that the CI/CD process lives up to its namesake.

Closing the loop

Figure 4.3 shows why the CI/CD methodology is effective for deploying incremental, reliable, and frequent software code changes into production:

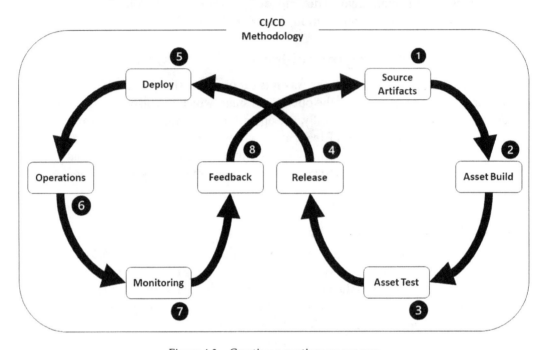

Figure 4.3 – Creating a continuous process

As *Figure 4.3* highlights, providing production feedback to developers, in essence, closes the loop, creating a continuous process whereby the developers can address the errors from the report, fix the source code, and update the artifact repository. Upon updating the artifact repository, a new release change of the CI/CD pipeline is triggered, resulting in the fixes being deployed into production.

So, as you can see, the CI/CD methodology inherently provides a continuous mechanism to deploy new software, software updates, or software fixes into production. Additionally, it should be evident that a successful implementation of the CI/CD pipeline requires multiple different teams, from software developers to infrastructure and IT teams.

This then begs the question, *would implementing a CI/CD methodology for ML address the deployment limitations highlighted at the outset?*

In the next section, we will answer this question, by exploring how the CI/CD methodology can be adapted to address an ML use case.

Automating ML with CI/CD

If you recall from *Chapter 1*, *Getting Started with Automated Machine Learning on AWS*, I highlighted that the typical ML process is manual and iterative. If you compare *Figure 1.2*, showing a realistic overview of the ML process, with *Figure 4.3*, showing the CI/CD process, I'm sure you will note that there are significant dissimilarities between the two processes:

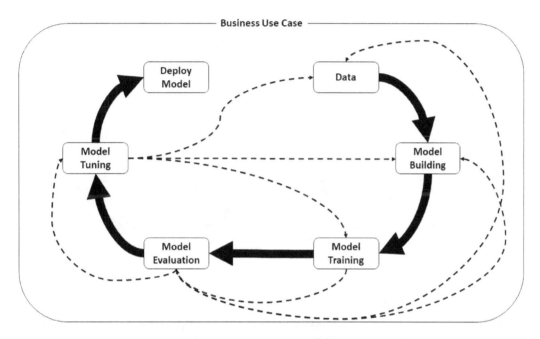

Figure 4.4 – A realistic overview of the ML process

However, since the focus of this chapter is to address the limitations of both the **typical** ML process and the AutoML methodology, specifically when it comes to bridging the gap for model deployment, there are several similarities between these processes. So, if you take a **deployment-centric** approach (*Figure 4.3*), as opposed to an **experiment-centric** approach (*Figure 1.2*), the procedure for deploying an optimized model into production is exactly the same as the procedure for deploying software code changes into production.

Taking a deployment-centric approach

Figure 4.5 shows what the resultant pipeline would look like if we were to take a deployment-centric approach to the ML process, using a software release CI/CD methodology:

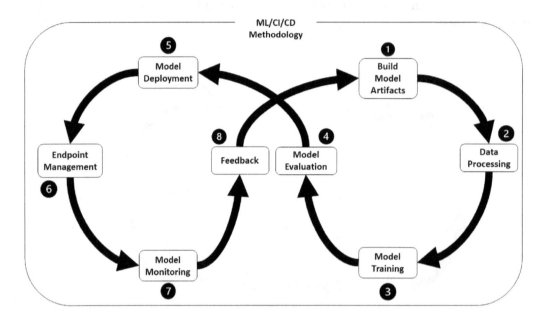

Figure 4.5 – Model deployment using CI/CD

As *Figure 4.5* shows, by treating a model deployment as a change release, we can automate the process using the CI/CD methodology. To further elaborate on exactly how this works, let's review each step of the process.

Building model artifacts

There are several components that can be considered model artifacts. For example, there is the algorithm code itself, as well as the various routines that leverage the algorithm code for training and evaluation. So, unlike the CI/CD pipeline that handles software code releases, there is no code compilation in the ML version of the CI/CD pipeline.

However, we can apply a similar methodology to the ones we used in *Chapter 3, Automating Complicated Model Development with AutoGluon*. By creating a container image with the relevant model artifacts, we can **compile** the image as a holistic model artifact. For example, if you refer to the *Building a deep learning container* section in *Chapter 3, Automating Complicated Model Development with AutoGluon*, you will recall that we created a `train.py` file to capture the model training and testing runtimes. We then built a `Dockerfile` deep learning container to capture the image build instructions so that we could use the `sm-docker-build` **CLI (command-line interface)** utility to compile the image as a holistic model artifact.

So, by storing the model artifacts in the pipeline repository, we can start the pipeline release cycle and compile or build a container image from the model artifacts.

Building data artifacts

Preparing training data is not an actual stage within the software release pipeline. However, if we view the task as building or compiling appropriate model training data and supplying a suitable runtime artifact to pre-process the data, then the data build task can then be thought of as a pipeline build task.

> **Note**
> You may recall from the previous section that a CI/CD release pipeline is triggered when code is added to or updated within a source code repository. This fact highlights a potential limitation in using the CI/CD methodology to deploy ML models. Since training data is not typically classified as source code, updating raw or training data won't trigger the release pipeline.

Building models

Training a model with the correct preprocessed data, as well as the correct parameters, can once again be viewed as building or compiling an optimized model. So, by changing our perspective on model training and optimization, and just like the data processing step, by supplying a suitable runtime artifact to execute the training process, we can treat the model training task as a pipeline build task.

Approving the model release

Just as with the software release pipeline, it follows that whether we are approving a software release for production or evaluating a trained model's performance, both of these tasks are in essence the same. In the case of a trained model, its performance is evaluated against the business criteria to determine whether or not it can be considered production-grade. If the model meets the business criteria, it can be released into production.

Deploying the model as a SageMaker endpoint

Once the model has been evaluated and approved for production, it can easily be deployed as a SageMaker hosted endpoint. SageMaker endpoints are essentially an endpoint address that can represent multiple models or, alternatively, multiple model versions (called variants). This translates to the fact that a SageMaker endpoint can inherently support a blue/green deployment strategy.

Releasing a new version of a trained model into an existing production endpoint means that SageMaker will automatically start redirecting new requests to the new model version while systematically phasing out the older model.

Therefore, incorporating SageMaker endpoints into the CI/CD pipeline provides the ML practitioner with the same deployment strategies as the software engineer.

Managing the SageMaker endpoint

Another compelling reason to deploy the released model as a SageMaker hosted endpoint is the fact that the SageMaker endpoint is an AWS managed service; therefore, there is no real need to manage any underlying operating system patches.

Additionally, hosted endpoints can be configured to automatically scale out as well as scale in, based on the number of usage requests.

Therefore, offloading the model deployment task to SageMaker significantly minimizes the operational overhead of managing the deployed model in production.

Monitoring the model's performance using Amazon SageMaker Model Monitor

Unlike the software release pipeline where logs, dashboards, and reports are used to provide feedback to developers, SageMaker endpoints can be incorporated into Amazon SageMaker Model Monitor to automatically verify that the production model is performing its intended purpose.

Amazon SageMaker Model Monitor statistically compares the responses from the production model against a baseline to automatically determine whether or not it is **drifting** from its intended purpose. If any of these constraint violations are detected, the ML practitioner can be alerted in order to address them as part of the next release of the pipeline, thus closing the feedback loop and making the entire deployment process continuous.

> **Note**
>
> Amazon SageMaker Model Monitor is capable of automating the monitoring tasks of the pipeline, provided endpoint data capture is enabled (https://docs.aws.amazon.com/sagemaker/latest/dg/model-monitor-data-capture.html) and a baseline is created (https://docs.aws.amazon.com/sagemaker/latest/dg/model-monitor-create-baseline.html).

As you can see, it is possible to use the CI/CD methodology to address the model deployment limitations highlighted at the outset, using a deployment-centric approach. But where do ML experiments fit into this methodology?

In the next section, we will explore how the process of finding the best model and its parameters can be incorporated into a CI/CD pipeline.

Creating an MLOps methodology

In the first section, *Introducing the CI/CD methodology*, I noted that software developers don't simply add new features or random updates to a code repository. They must perform a unit test to ensure that updates are functional before deploying changes into production.

This outcome corresponds to the overall objective behind performing an ML experiment. The goal behind the ML experiment is to get the best candidate model, with its associated parameters, before deploying it into production. Integrating the ML experiment into the **development** and **operate** methodology of **DevOps** is the basis of an MLOps methodology; *Figure 4.6* provides an overview of the process:

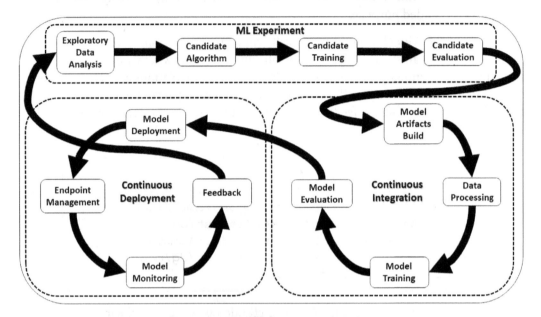

Figure 4.6 – An MLOps process overview

Figure 4.6 clearly shows how an ML practitioner can take a two-phased approach to automating the ML process using the MLOps methodology. By using an AutoML methodology to automate the ML experiment, and generate the best candidate artifacts, the resultant artifacts can be submitted into the source repository to trigger a production-grade deployment using the CI/CD pipeline.

So, now we have the necessary background on how the CI/CD methodology functions and how to integrate ML into the process, we can now apply these techniques to the *age calculator* use case. However, before diving into a hands-on example, in the next section, we are going to review the various capabilities that AWS provides to create a CI/CD pipeline.

Creating a CI/CD pipeline on AWS

AWS provides an entire suite of developer tools that address the many requirements for hosting code, and building and deploying pipeline assets. To create a CI/CD pipeline on AWS, we will be making use of three primary services within the AWS developer toolchain. To further simplify the construction and automation of the pipeline, we will make use of two additional services within the AWS development suite.

> **Note**
>
> To learn more about the development tools available from AWS, you can reference the product page (`https://aws.amazon.com/products/developer-tools/`).

Let's review the important services that make up the CI/CD pipeline.

Using the AWS CI/CD toolchain

The three core components that make up a CI/CD pipeline are as follows:

- A component to store the various pipeline artifacts
- A component process to build the various pipeline assets
- A component to automate the pipeline execution

To facilitate creating these three core components, AWS provides dedicated services to match the required capabilities of each component, namely the following:

- **AWS CodeCommit**
- **AWS CodeBuild**
- **AWS CodePipeline**

While there are other CI/CD pipeline components and associated AWS services, we will be focusing on these three to provide the necessary capabilities for ML release automation.

AWS CodeCommit

CodeCommit (`https://aws.amazon.com/codecommit/`) is a cloud-based source code and version control service. In essence, it is the AWS-managed alternative to GitHub. We will be using CodeCommit to store all the various pipeline and ML model artifacts.

> **Note**
>
> If you are unfamiliar with the concept of source code and version control, AWS provides an overview on their website (`https://aws.amazon.com/devops/source-control/`).

AWS CodeBuild

CodeBuild (`https://aws.amazon.com/codebuild/`) is the heart of the continuous integration phase of the CI/CD pipeline. This service is responsible for compiling or building the various artifacts into usable pipeline assets. In the case of ML release automation, CodeBuild builds the required model training and serves runtimes, as well as executing the data processing, model training, and model evaluation processes.

AWS CodePipeline

CodePipeline (`https://aws.amazon.com/codepipeline/`) handles the continuous deployment phase of the CI/CD pipeline. This service contains the structure of the pipeline and is responsible for automating the task of releasing the trained model into production as a SageMaker hosted endpoint.

Now that we have a brief overview of the core AWS services and their purposes, in the next section, I'll highlight some additional AWS capabilities to build the service infrastructure.

Working with additional AWS developer tools

We will be making use of two additional AWS developer tools, and while these services are not critical to the success of a CI/CD implementation, they make it easier to develop an entire solution. For example, we will use the AWS **Cloud Development Kit (CDK)** to codify the entire solution. Thus, not only are the pipeline artifacts sourced and managed as code but also the pipeline itself and the associated AWS infrastructure. This makes the entire solution a cloud-native application.

> **Note**
>
> You will see in later chapters how creating the entire solution as a cloud-native application can further streamline the end-to-end process for ML automation. As you will see, the CDK framework (`https://aws.amazon.com/cdk/`) will play a fundamental part in enhancing the process. If you are new to the notion of using the CDK to codify a cloud-native solution, it is highly recommended that you review the CDK documentation (`https://docs.aws.amazon.com/cdk/api/latest/`), refer to the samples in the CDK GitHub repository (`https://github.com/aws-samples/aws-cdk-examples`), and look at the official AWS CDK workshop (`https://cdkworkshop.com/`).

To ensure consistency and ease of use for the hands-on example, we will make use of AWS's cloud-based IDE service called **Cloud9** (`https://aws.amazon.com/cloud9/`). While it is possible to run through the hands-on example with a local IDE, Cloud9 has all the associated tools, programming libraries, and utilities pre-installed.

We now have an overview of the CI/CD process, how it can be applied to automatically release ML models into production, and the AWS services we can use to build the solution as a cloud-native application. So, let's apply what we've learned to the *age calculator* use case by means of a hands-on example.

Creating a cloud-native CI/CD pipeline for a production ML model

As a guide to successfully implementing a CI/CD pipeline for the *age calculator* use case, we will be performing the following tasks:

1. Preparing the development environment
2. Creating the pipeline artifact repository
3. Developing the application artifacts
4. Deploying the pipeline application
5. Creating the ML model artifacts
6. Executing the automated ML model deployment

Furthermore, as we build out the solution, using the previous steps, I will break the various tasks down into two separate categories. One category will center around the various tasks performed by the application development team, and the other category will encompass the tasks that are typically performed by the ML practitioner. My primary objective in doing this is to highlight the roles of a cross-functional team.

Effective coordination between the ML practitioners and the DevOps engineering team establishes the fundamental foundation for successful model deployment. This process of working together at a foundational level, thus establishing a cross-functional team, is the primary success criteria for successful model deployment.

By the end of this section, you will see that it's not only the toolchain used or even the execution of the pipeline itself that determines the successful implementation of a production-grade ML model; rather, the key element is a cross-functional team.

Now, let's get started by creating the Cloud9 development environment.

> **Tip**
>
> Reference files for the following code examples can be found in the companion GitHub repository (`https://github.com/PacktPublishing/Automated-Machine-Learning-on-AWS/tree/main/Chapter04/`) for this chapter.

Preparing the development environment

To begin, we will start by looking at this undertaking from the perspective of the DevOps engineer, by preparing the Cloud9 environment for application development. The following steps will walk you through this process:

1. Log into the AWS account you've been using and open the Cloud9 management console (`https://console.aws.amazon.com/cloud9`) for your supported AWS Region.

2. Create a Cloud9 environment by clicking the **Create environment** button.

3. When prompted, provide a name and an optional description. Then, click the **Next step** button. *Figure 4.7* shows an example of naming the environment:

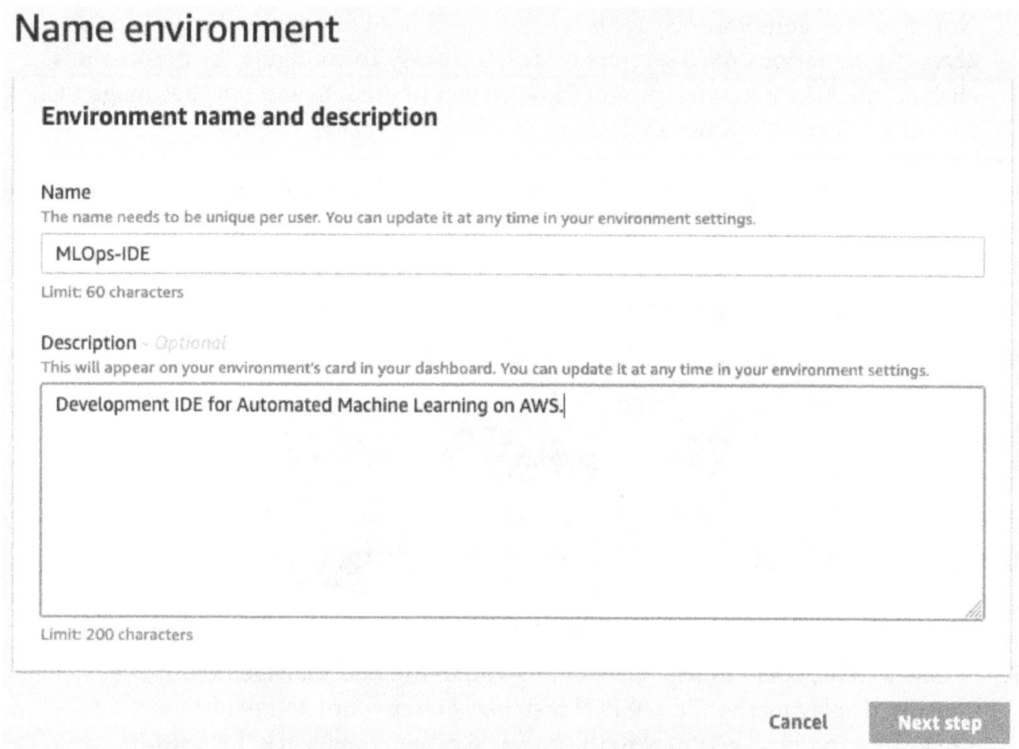

Name environment

Environment name and description

Name

The name needs to be unique per user. You can update it at any time in your environment settings.

MLOps-IDE

Limit: 60 characters

Description - *Optional*

This will appear on your environment's card in your dashboard. You can update it at any time in your environment settings.

Development IDE for Automated Machine Learning on AWS.

Limit: 200 characters

Cancel   Next step

Figure 4.7 – Naming the Cloud9 environment

4.   On the **Configure settings** page, accept the default **Environment settings** by
clicking the **Next step** button.

> **Note**
> Accepting the default settings will ensure that the Cloud9 environment is
> eligible for the AWS Free Tier. However, it is recommended that you use a
> t3.small instance, which is not eligible for the Free Tier.

5.   On the **Review** page, confirm the settings and click **Create environment**. After
a few minutes, you will be redirected to the IDE web interface.

> **Tip**
> To familiarize yourself with the IDE and how to use the various panels, review
> the basic tour documentation on the AWS website (https://docs.aws.
> amazon.com/cloud9/latest/user-guide/tutorial-tour-
> ide.html).

6. Now that the Cloud9 workspace is ready, we will need to provide the appropriate access to the various AWS services we will be using. To configure the permissions, click on the **A** icon in the top right-hand corner of the IDE and select **Manage EC2 Instance**. *Figure 4.8* shows an example of what the process looks like:

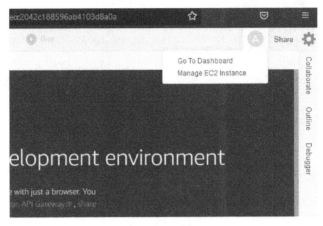

Figure 4.8 – Manage EC2 Instance

7. A new web browser tab will open, taking you to the EC2 management console, displaying the Cloud9 EC2 instance. Select your Cloud9 instance by clicking the checkbox next to the instance name. Then, click the **Actions** button, and from the drop-down menus, select the **Security** option. After the security menu expands, select **Modify IAM role**. *Figure 4.9* shows an example of the expanded menu settings:

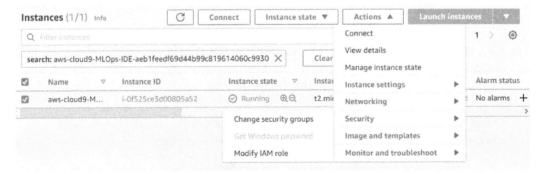

Figure 4.9 – The EC2 instance security menus

8. When the **Modify IAM role** page opens, click the **Create new IAM role** link to open the IAM management console in a new browser tab. *Figure 4.10* shows an example of the **Modify IAM role** page:

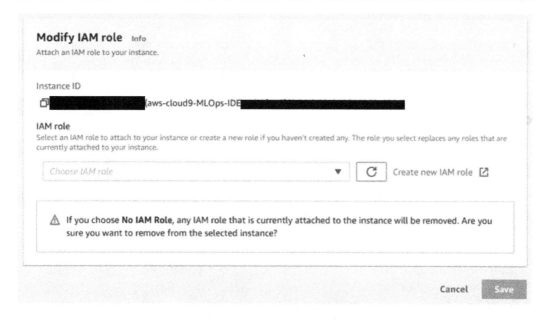

Figure 4.10 – Modify IAM role

9. Within the IAM management console, click the **Create role** button to create a new instance administrator role.

10. On the **Create role** page, select **EC2**, under the **Common use cases** section, and then click the **Next: Permissions** button. *Figure 4.11* shows an example of selecting the **EC2** use case:

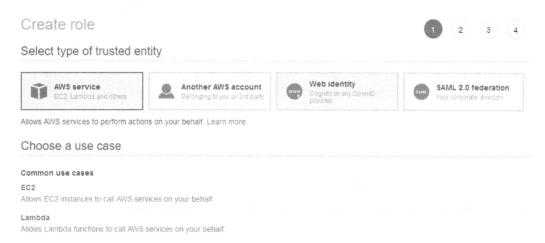

Figure 4.11 – The EC2 common use case

11. Using the provided search bar in the **Attach permissions policies** section, enter `administrator` as the search term. You will see the various policies containing `administrator` listed. Select the checkbox next to **AdministratorAccess** and then click the **Next: Tags** button. *Figure 4.12* shows an example of selecting the **AdministratorAccess** policy:

Figure 4.12 – Selecting the AdministratorAccess policy

12. Skip the **Add tags (optional)** section by clicking on the **Next: Review** button.

13. On the **Review** page, enter an appropriate role name and click the **Create role** button. *Figure 4.13* shows an example of providing a role name:

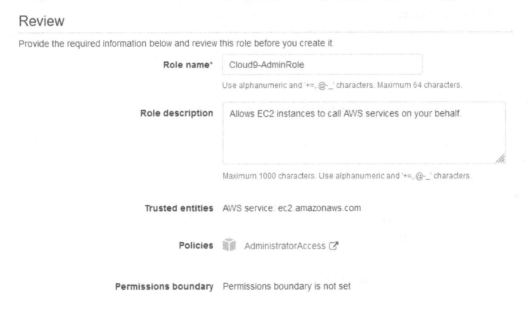

Figure 4.13 – Providing a role name

14. Once the role has been created, you can close the IAM console and return to the **Modify IAM role** tab from *step 8*.

15. Click the refresh icon and, using the dropdown, select the role you created in *step 13*. *Figure 4.14* shows an example of what the page looks like:

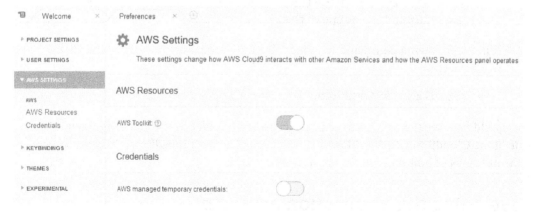

Modify IAM role   Info

Attach an IAM role to your instance.

Instance ID

▢ ▮▮▮▮▮▮▮▮ (aws-cloud9-MLOps-IDE▮▮▮▮▮▮▮▮▮▮▮▮▮▮▮▮▮)

IAM role

Select an IAM role to attach to your instance or create a new role if you haven't created any. The role you select replaces any roles that are currently attached to your instance.

| Cloud9-AdminRole | ▼ | ↻ | Create new IAM role ⎘ |

Cancel   Save

Figure 4.14 – Selecting the IAM role

16. Click the **Save** button.

17. Go back to the browser tab displaying the Cloud9 workspace and attach the newly created role by clicking on the gear icon in the top right-hand corner.

18. In the workspace **Preferences** tab, select the **AWS SETTINGS** option and disable the **AWS managed temporary credentials** switch. *Figure 4.15* shows an example of what the final **AWS Settings** page will look like:

Figure 4.15 – Disabling AWS managed temporary credentials

Now that the development environment has been set up, we can proceed to the next task of creating the pipeline artifact repository.

Creating the pipeline artifact repository

Using the development environment, we will now create a CodeCommit repository to store the various pipelines and, eventually, the ML model artifacts. Although there are multiple ways to create a CodeCommit repository, we will be using the AWS CLI, which is already installed and configured in the Cloud9 workspace. The following steps will walk us through this process:

1. Using the terminal pane (the bottom section of the Cloud9 workspace), run the following CLI command to ensure that the CLI region settings are correct. Make sure to replace <REGION> with the AWS Region you are currently using:

    ```
    $ aws configure set region <REGION>
    ```

2. Create a CodeCommit repository called abalone-cicd-pipeline, using the following command:

    ```
    $ aws codecommit create-repository --repository-name
    abalone-cicd-pipeline --repository-description "Automated
    ML on AWS using CI/CD"
    ```

3. Next, we capture the URL for the newly created repository in order to clone it. Run the following command to create the CLONE_URL parameter:

    ```
    $ CLONE_URL=$(aws codecommit get-repository --repository-
    name abalone-cicd-pipeline --query "repositoryMetadata.
    cloneUrlHttp" --output text)
    ```

4. Run the following command to clone the empty repository, locally, in the Cloud9 workspace:

    ```
    $ git clone $CLONE_URL
    ```

You should now see the abalone-cicd-pipeline folder in the left-hand navigation pane of the Cloud9 workspace. Now that we have our project repository, we can proceed to the next task of building out the application artifacts.

Developing the application artifacts

Before we can start codifying the entire solution, we need to configure the application environment. The next set of steps will configure the environment to use the AWS CDK.

Creating and configuring the CDK project

If you refer to the CDK documentation (`https://docs.aws.amazon.com/cdk/latest/guide/getting_started.html`), there are certain prerequisites that need to be configured before using the CDK. Fortunately, AWS assists with these prerequisites by pre-configuring them within the Cloud9 IDE. So, all we need to do before building out the application is to update to the latest version of the CDK and set up the environmental variables, following these steps:

> **Note**
>
> At the time of writing, the latest version of the AWS CDK is **2.3.0 (build beaa5b2)**. In order to maintain the functionality of the code within this example, we will use version 2.3.0 of the CDK:

1. Before building the codified CDK application, run the following command to ensure that we have a consistent version of the CDK installed:

   ```
   $ npm install -g aws-cdk@2.3.0 --force
   ```

2. Run the following command to confirm that version 2.3.0 is the current version of the CDK:

   ```
   $ cdk --version
   ```

   > **Note**
   >
   > Make sure to remember the version of the CDK, as this information will be required in a later step.

3. Next, we run the following set of commands to configure some of the CDK environment variables, such as our AWS account and the AWS Region we are currently using:

   ```
   $ export CDK_DEFAULT_ACCOUNT=$(aws sts get-caller-identity --query "Account" --output text)
   $ echo "export CDK_DEFAULT_ACCOUNT=$(aws sts get-caller-identity --query "Account" --output text)" >> ~/.bashrc
   $ export CDK_DEFAULT_REGION=$(aws configure get region)
   $ echo "export CDK_DEFAULT_REGION=$(aws configure get region)" >> ~/.bashrc
   ```

4. Create an empty CDK project and specify Python as the project's programming language by running the following command:

```
$ cd abalone-cicd-pipeline && cdk init --language python
```

5. Since the CDK Python project will interface with the artifact repository, we can create the primary branch for the project using the following commands:

```
$ git add -A
$ git commit -m "Started CDK Project"
$ git branch main
$ git checkout main
```

6. Next, we can configure the Python environment by running the following commands:

```
$ source .venv/bin/activate
$ python -m pip install --upgrade pip pylint boto3
$ pip install -r requirements.txt
```

With the CDK project created and configured, we can now move on to building the application artifacts.

Creating the application

Now that we have prepared the CDK project environment, it's at this stage of the process that cross-team collaboration becomes crucial to the continued success of the project. We, as the application developers, now need to work with the ML practitioner team to assess the following key elements of the application:

- We need to understand what the final applications will look like. In this case, the final application will be a production-grade ML model, deployed as a SageMaker hosted endpoint.

- We also need to understand what the ML practitioner team will be contributing as their pipeline artifacts. In this case, the ML practitioner team will deliver a customer SageMaker container image, such as the container images we worked with in *Chapter 3, Automating Complicated Model Development with AutoGluon*.

- We will need to understand how to build or compile these artifacts. In essence, we need to understand what the build runtime logic will entail. In this case, the ML practitioners will want to use SageMaker to process the training data, train the ML model, and evaluate its performance against the business criteria for the use case.

- We will need to understand what dependencies are required by the ML practitioner artifacts. For instance, the model data processing and training components will require access to the raw source data.

- Just as important, we need to assess what security and access requirements are needed by the relevant AWS services, as well as the various teams creating and updating the application artifacts.

Once we have captured, reviewed, and all team members have signed off on these requirements, we can go ahead and build out the application. The first part of the overall application we are going to develop is the final piece, the SageMaker hosted endpoint.

> **Note**
>
> It may seem counterintuitive to start the application development process by focusing on the final piece of the pipeline – in this case, the production-grade model. In most situations, it is a good practice to start the development of an automated workflow by focusing on the outcome. This way, you can work backward, from the end result, to understand and develop the necessary code that eventually produces the final outcome.

Codifying the SageMaker endpoint

Since this may be your first time working with the CDK, code for the different constructs has already been provided for you in the companion GitHub repository for this chapter. Use the following steps to add the SageMaker endpoint construct into the CDK environment:

1. Using the terminal windows within the Cloud9 workspace, run the following command to clone the companion GitHub repository:

```
$ cd ~/environment/ && git clone https://github.com/
PacktPublishing/Automated-Machine-Learning-on-AWS src
```

2. Copy the pre-built abalone_endpoint_stack.py file into the abalone_cicd_pipeline folder with the following commands:

```
$ cd ~/environment/abalone-cicd-pipeline
$ cp ~/environment/src/Chapter04/cdk/abalone_endpoint_
stack.py abalone_cicd_pipeline/
```

3. Using the left-hand navigation panel of the Cloud9 workspace, expand the `abalone-cicd-pipeline` folder, and then expand the `abalone_cicd_pipeline` folder to reveal the `abalone_endpoint_stack.py` file.

4. Double-click on the `abalone_endpoint_stack.py` file so that we can review the code.

Now that the `abalone_endpoint_stack.py` file is open in the Cloud9 editor, we can walk through the code to review how we build the hosted endpoint. The first thing you will see once opening the file is that we need to import the necessary CDK modules for our construct and the `aws_sagemaker` modules. We then initialize a Python class for `EndpointStack()` as a `cdk.Stack` construct; thus, we are essentially instantiating a new CloudFormation stack with the relevant SageMaker endpoint resources.

> **Note**
>
> If you are unfamiliar with what a CloudFormation stack is, or how the CDK initializes the AWS resources as components of the stack construct, you can refer to the AWS documentation for stacks (`https://docs.aws.amazon.com/AWSCloudFormation/latest/UserGuide/stacks.html`) and constructs (`https://docs.aws.amazon.com/cdk/latest/guide/constructs.html`).

Next, we define parameters for the CloudFormation stack, such as the name of the S3 bucket housing our data or the `bucket_name` parameter. As you will see later in this chapter, these parameters will be supplied to the stack as outputs from a pipeline execution.

After declaring the various CloudFormation stack parameters, we instantiate a representation of the trained model, using the `CfnModel()` module from the `aws_sagemaker` library. Here, we define the necessary parameters to create to tell SageMaker about the trained model so that it can be hosted as a SageMaker endpoint.

> **Note**
>
> For more information on the different parameters required to represent a trained model, you can refer to the `CfnModel` documentation (`https://docs.aws.amazon.com/cdk/api/latest/python/aws_cdk.aws_sagemaker/CfnModel.html`).

After defining the model, we specify the necessary configuration parameters required to actually host the trained model. This is done using the `CfnEndpointConfig()` module from the `aws_sagemaker` library. Here, we define the type and amount of compute resources to host the model. You can see that we also specify the `data_capture_config` parameter to tell SageMaker where to store the inference request payload coming into the hosted model, as well as the inference response output coming from the hosted model. This way, we are essentially logging the endpoint usage so that we can monitor the model in production.

Lastly, we define the endpoint itself, using the `CfnEndpoint()` module. Here, we define a name for the endpoint and specify the endpoint configuration to use.

Next, we will build the runtime logic to retrieve the specific `CfnParamater()` values from the pipeline execution.

Configuring the deployment parameters

In order to provide the required `CfnParamater()` parameters to the `EndpointStack()` construct, as shown in the previous steps, we need to capture and store the pipeline execution parameters in a JSON file called `params.json`. You will see that once we define the actual pipeline construct, this file then serves as the input to the endpoint CloudFormation stack. The following steps show you how to copy the runtime script for review:

1. Using the terminal within the Cloud9 workspace, create a folder to store the pre-built scripts by running the following command:

    ```
    $ cd ~/environment/abalone-cicd-pipeline/
    $ mkdir -p artifacts/scripts/
    $ cp ~/environment/src/Chapter04/scripts/deploy.py
    artifacts/scripts/
    ```

2. Using the left-hand navigation panel of the Cloud9 workspace, expand the newly created `artifacts` folder.

3. Now, expand the `scripts` folder and double-click on `deploy.py` for review.

With the `deploy.py` file open in the Cloud9 editor panel, we can review just how to get the execution parameters from a running pipeline and create the `params.json` file.

After importing the necessary Python libraries for the script, you will see from the following code snippet that we configure the AWS SDK for Python to access both the SageMaker and CodePipeline SDK clients after we've set up logging and configured our global environment parameters:

```
...
logger = logging.getLogger()
logging_format = "%(levelname)s: [%(filename)s:%(lineno)s]
%(message)s"
logging.basicConfig(format=logging_format, level=os.environ.
get("LOGLEVEL", "INFO").upper())
codepipeline_client = boto3.client("codepipeline")
sagemaker_client = boto3.client("sagemaker")
pipeline_name = os.environ["PIPELINE_NAME"]
model_name = os.environ["MODEL_NAME"]
role_arn = os.environ["ROLE_ARN"]
...
```

Setting up logging is important for us to verify how the script is being executed and ensure that it's functioning correctly, and if not, logging errors will allow us to troubleshoot and debug.

Next, we've created two Python functions, namely the get_execution_id() and get_model_artifact() functions. These functions are used in the __main__ program to get the unique pipeline execution ID from CodePipeline, as well as the name of the trained ML model from the SageMaker model registry.

The __main__ program then takes the parameters returned by both the get_execution_id() and get_model_artifact() functions to populate the params.json file. We will use the pipeline execution ID for asset versioning. As you will see later, we will append this ID to the various assets, specific to the release, in order to track the model's lineage from source to the release.

Now that we have the necessary Python code to query the pipeline, and have retrieved the necessary execution parameters to supply them to the deployment construct, we have essentially created the necessary artifacts we need to run the continuous deployment phase of the pipeline. Next, we can work on the artifacts required by the continuous integration phase of the pipeline.

Configuring the build artifacts

As we continue with the backward-working methodology, in this next step, we are going to create the artifacts needed to build, train, and evaluate the ML model. The following steps will walk you through how to do this:

1. Using the terminal within the Cloud9 workspace, run the following command to copy the pre-built `build.py` script into the `scripts` folder:

    ```
    $ cd ~/environment/abalone-cicd-pipeline/
    $ cp ~/environment/src/Chapter04/scripts/build.py
    artifacts/scripts/
    ```

2. Using the left-hand navigation panel of the Cloud9 workspace, double-click on `build.py` for review.

As you saw with the `build.py` file, the `deploy.py` script imports the necessary Python libraries, sets up logging, and also defines the same `get_execution_id()` and `get_model_artifact()` Python functions. We also create specific Python functions to initiate the appropriate stage of the ML process. For example, to train the ML model, we call the `handle_training()` function. This function makes the necessary API call to SageMaker to start the training job. As you can see, we apply the same approach to both the `handle_data()` function to preprocess the training and validation datasets, as well as the `handle_evluation()` function to evaluate the trained ML model's performance.

We also create a new function called `handle_status()`, which acts as a wrapper for each step of the ML process. The following code snippet shows the `handle_status()` Python function:

```
. . .
def handle_status(task=None, job_name=None):
    if task == "preprocess" or task == "evaluate":
        status = sagemaker_client.describe_processing_
job(ProcessingJobName=job_name)["ProcessingJobStatus"]
        while status == "InProgress":
            time.sleep(60)
            logger.info(f"Task: {task}, Status: {status}")
            status = sagemaker_client.describe_processing_
job(ProcessingJobName=job_name)["ProcessingJobStatus"]
        return status
    elif task == "train":
```

```
        status = sagemaker_client.describe_training_
job(TrainingJobName=job_name)["TrainingJobStatus"]
        while status == "InProgress":
            time.sleep(60)
            logger.info(f"Task: {task}, Status: {status}")
            status = sagemaker_client.describe_training_
job(TrainingJobName=job_name)["TrainingJobStatus"]
        return status
...
```

As you can see from the code snippet, depending on the current stage of the pipeline's execution, denoted by the `task` parameter, the `handle_status()` function will call the appropriate handle function to get the status of the SageMaker job associated with the particular task or stage of the ML process. For example, to train the ML model, the `handle_status()` function determines from the `task` parameter that it needs to get the status of the SageMaker training job and log whether the task it has is currently running or in progress.

Finally, we have the __main__ function, shown in the following snippet, as the primary execution point for the script:

```
...
if __name__ == "__main__":
    task = sys.argv[1]
    execution_id = get_execution_id(name=pipeline_name,
task=task)
    logger.info(f"Executing {task.upper()} task")
    if task == "preprocess":
        job_name = handle_data(model_name=model_name,
execution_id=execution_id)
        status = handle_status(task=task, job_name=job_name)
    elif task == "train":
        job_name = handle_training(model_name=model_name,
execution_id=execution_id)
        status = handle_status(task=task, job_name=job_name)
    elif task == "evaluate":
        job_name = handle_evaluation(model_name=model_name,
execution_id=execution_id)
        status = handle_status(task=task, job_name=job_name)
    else:
```

```
        error = "Invalid argument: Specify 'preprocess',
'train' or 'evaluate'"
        logger.error(error)
        sys.exit(255)
    if status == "Completed":
        logger.info(f"Task: {task}, Final Status: {status}")
        sys.exit(0)
    else:
        error = f"Task: {task}, Failed! See CloudWatch Logs for
further information"
        logger.error(error)
        sys.exit(255)
...
```

As you can see from the preceding code snippet, the __main__ function takes the current pipeline stage as input and calls the appropriate handler function for that stage. For example, if the pipeline is currently executing the training stage, the __main__ function determines from the task input that it needs to call the handle_training() function to initiate the SageMaker training job, and then the handle_status() function to track and manage the execution of that training job.

> **Note**
> Refer to the AWS SDK for Python documentation for more information on the various parameters to create a SageMaker processing job (https://boto3.amazonaws.com/v1/documentation/api/latest/reference/services/sagemaker.html#SageMaker.Client.create_processing_job) and SageMaker training job (https://boto3.amazonaws.com/v1/documentation/api/latest/reference/services/sagemaker.html#SageMaker.Client.create_training_job).

While it may not seem inherently intuitive at this point, by creating both the build.py, and deploy.py scripts, we have just produced the fundamental mechanisms by which the pipeline will execute the continuous integration and continuous deployment process. For example, by executing the build.py script, the pipeline can build, train, and evaluate a production-grade ML model. And, by executing the deploy.py script, the pipeline can process the relevant parameters from the integration to deploy the model into production by means of the endpoint CDK construct.

However, before moving on to the next chapter, it is a good practice to commit the current work in progress to the source code repository. The following steps will walk you through how to checkpoint your progress:

1. Using the terminal within the Cloud9 workspace, run the following commands to add the changes we've made to the working directory:

    ```
    $ cd ~/environment/abalone-cicd-pipeline/
    $ git add -A
    ```

2. Now, run the command to commit these changes to the repository history:

    ```
    $ git commit -m "Checkpoint"
    ```

3. Finally, we can push the changes to the source code repository by running the following command:

    ```
    $ git push --set-upstream origin main
    ```

So, now that these intrinsic artifacts have been created and committed to the repository, we can continue to develop the pipeline itself in the next chapter.

Summary

In this chapter, you were introduced to the concept of a CI/CD process as a way to close the gap between building a production-grade ML and getting the model into production. Making use of this methodology, an ML practitioner doesn't simply hand over the trained model to the platform teams but rather integrates the model artifacts into the overall process.

While we haven't as yet shown how the ML practitioner contributes these model artifacts into a process, we have established a pattern of codifying the process by introducing and setting up an AWS CDK project. By using the CDK, we practically demonstrated a backward-working approach for how the engineering team can deploy a trained model as a SageMaker-hosted endpoint CDK construct. We also demonstrated how the engineering teams built the fundamental mechanisms that will eventually automate the integration of the model training and evaluation procedures into the process.

In the next chapter, we will continue building out the CI/CD pipeline, adding the model artifacts and automatically deploying the trained model into production.

5
Continuous Deployment of a Production ML Model

In *Chapter 4, Continuous Integration and Continuous Delivery (CI/CD) for Machine Learning*, we were introduced to the concept of continuous integration, and continuous deployment, as a means of bridging the gap between ML model development and ML model deployment. We were also introduced to the AWS CDK, as a way to further close this gap, by bringing the different artifacts that software engineers and ML practitioners develop into a single cloud-native application. Thus, allowing us to codify a CI/CD pipeline that automates the entirety of the ML process. Closing this gap, and helping to facilitate this inter-team synergy, is one of the core design philosophies behind why AWS originally created the CDK.

> **Note**
>
> For more information on the AWS CDK philosophy, you can read the best practices for developing cloud applications in the AWS CDK blog post (`https://aws.amazon.com/blogs/devops/best-practices-for-developing-cloud-applications-with-aws-cdk/`).

Although we started creating the core mechanisms for training and deploying the ML model, we have yet to create the overall pipeline, responsible for orchestrating the process. In this chapter, we will pick up from where we left off, by continuing to codify the CI/CD pipeline construct, as well as the ML model artifacts. The following topics will emphasize how we will accomplish these tasks:

- Deploying the CI/CD pipeline
- Building the ML model artifacts
- Executing the CI/CD pipeline

Technical requirements

This chapter will use the following resources:

- A web browser (for the best experience, it is recommended that you use the Chrome or Firefox browser).
- Access to the AWS account that you've been using in the previous chapters.
- Access to the Cloud9 IDE that you used to start building the CDK application in *Chapter 4, Continuous Integration and Continuous Delivery (CI/CD) for Machine Learning*.
- Access to the same SageMaker Studio UI we used in *Chapter 3, Automating Complicated Model Development with AutoGluon*.
- We will once again be working within the usage limits of the AWS Free Tier to avoid exceeding unnecessary costs.
- Source code samples for the CDK constructs, and ML model artifacts, are provided in the companion GitHub repository for this chapter (`https://github.com/PacktPublishing/Automated-Machine-Learning-on-AWS/tree/main/Chapter05`).

Deploying the CI/CD pipeline

You will recall from *Chapter 4, Continuous Integration and Continuous Delivery (CI/CD) for Machine Learning*, that we concluded the chapter by checkpointing the intrinsic artifacts, namely the `buld.py` and `deploy.py` scripts, and committing them into the CodeCommit repository. Whereas these artifacts fundamentally create and deploy a trained ML model, we still need to wrap them in a continuous integration and continuous deployment process. To accomplish this, we will continue using the AWS CDK to create a codified CI/CD pipeline construct.

Codifying the pipeline construct

The penultimate component of the application is the pipeline construct itself. Using the following steps, we will once again leverage the AWS CDK to create the pipeline:

1. If you don't already have the Cloud9 environment open in your web browser, log into the AWS account you've been using, and open the Cloud9 management console (`https://console.aws.amazon.com/cloud9`) for your AWS region. Click on the **Open IDE** button to launch the Cloud9 instance. Once the Cloud9 instance is online, use the **Terminal** panel to activate the Python virtual environment, by running the following commands:

    ```
    $ cd ~/environment/abalone-cicd-pipeline/
    $ source .venv/bin/activate
    ```

2. Now, run the following command to add the pre-built pipeline construct, `abalone_cicd_pipeline_stack.py`, into the `abalone_cicd_pipeline` folder:

    ```
    $ cp ~/environment/src/Chapter05/cdk/abalone_cicd_
    pipeline_stack.py ~/environment/abalone-cicd-pipeline/
    abalone_cicd_pipeline/
    ```

> **Note**
>
> When we initialized the CDK application in the previous chapter, a templatized `abalone_cicd_pipeline_stack.py` construct was created for you. We will be replacing this file with an updated version, pre-built for our example. If you have not already cloned the companion GitHub repository, you can refer to the *Codifying the SageMaker Endpoint* section in *Chapter 4, Continuous Integration and Continuous Delivery (CI/CD) for Machine Learning*.

3. Using the left-hand navigation panel of the Cloud9 workspace, double-click on the updated version of the `abalone_cicd_pipeline_stack.py` file for review.

The first thing you will note as we walk through the code is that we import the necessary CDK modules we will be using to create the construct resources. The primary modules we will be using within this contract are the `aws_codepipeline`, `aws_codpipeline_actions`, `aws_codebuild`, and `aws_iam` modules.

Next, as you can see from the following code snippet, we define the `PipelineStack()` class, as `cdk.Stack`, and initialize it:

```
. . .
class PipelineStack(cdk.Stack):
    def __init__(self, scope: Construct, id: str, *, model_
name: str=None, repo_name: str=None, cdk_version: str=None,
**kwargs) -> None:
        super().__init__(scope, id, **kwargs)
. . .
```

As you see from the previous code snippet, we also supply some key parameters, namely `model_name`, `repo_name`, and `cdk_version` to initialize the class. These parameters are specific to our CDK application and will be defined later in this chapter when we instantiate the CDK application itself.

Once we've initialized the construct, the first resource we need to create is `sagemaker_role`. This is an IAM role that the runtime logic scripts, namely `build.py` and `deploy.py`, will assume to execute the various SageMaker tasks. For example, `sagemaker_role` has **FullAccess** to the SageMaker service, in order to process the training data, train, evaluate and deploy the model.

Next, we define variables for the repositories that will contain the artifact source code. For example, we define a variable called `container_repo` to declare the CodeCommit repository, as well as the variable called `s3_bucket` where the raw training data and pipeline execution artifacts will be stored.

Next, we define the first of four CodeBuild projects. These CodeBuild projects execute the runtime logic to essentially build the required pipeline assets that preprocess the data, train the ML model, evaluate the ML model, and construct the deployment parameters needed to deploy the model. For example, and as shown in the following code snippet, the `container_build` project takes the ML model artifact and executes the runtime logic to build and store the artifact as a Docker image:

```
...
                    build=dict(
                        commands=[
                            "echo Build started on `date`",
                            "echo Building the Docker
image...",
                            "docker build -t $IMAGE_REPO_
NAME:$IMAGE_TAG --build-arg REGION=$AWS_DEFAULT_REGION ."
                        ]
                    ),
                    post_build=dict(
                        commands=[
                            "echo Build completed on
`date`",
                            "echo Pushing the Docker
image...",
                            "docker push $IMAGE_REPO_
NAME:$IMAGE_TAG"
                        ]
                    )
...
```

Since we will be basing our Docker container image on the AWS Deep Learning Containers, using the same methodology from *Chapter 3, Automating Complicated Model Development with AutoGluon*, we also need to provide the CodeBuild project with the necessary permissions to access the container repositories. You can see from the following code snippet that we add an IAM `PolicyStatement()` to the CodeBuild project, giving the IAM role access to the DLC container repositories:

```
...
        container_build.role.add_to_policy(
            iam.PolicyStatement(
                resources=[
```

```
                          "arn:aws:ecr:*:763104351884:repository/*",
                          "arn:aws:ecr:*:217643126080:repository/*",
                          "arn:aws:ecr:*:727897471807:repository/*",
                          "arn:aws:ecr:*:626614931356:repository/*",
                          "arn:aws:ecr:*:683313688378:repository/*",
                          "arn:aws:ecr:*:520713654638:repository/*",
                          "arn:aws:ecr:*:462105765813:repository/*"
                  ],
                  actions=[
                          "ecr:BatchGetImage",
                          "ecr:GetDownloadUrlForLayer"
                  ],
                  effect=iam.Effect.ALLOW
          )
        )
  ...
```

The next CodeBuild project we define, called `data_build`, executes the runtime logic for the data processing task. As you can see from the following code snippet, we run the previously created `build.py` script, and supply the `preprocess` argument, telling the Python script to make an API call for SageMaker to run the processing Job:

```
  ...
                  "build": {
                          "commands": [
                                  "echo Build started on `date`",
                                  "python ./artifacts/scripts/
build.py preprocess"
                          ]
                  },
  ...
```

The next two CodeBuild projects, namely `model_build` and `evaluation_build`, execute the same runtime logic as the `data_build` project. Except that `model_build` supplies the `train` parameter to the `build.py` script to make an API call for SageMaker to run the ML model training job. For example, you can see the following code snippet, where the `train` parameter is supplied to the CodeBuild project:

```
...
            "build": {
                "commands": [
                        "echo Build started on `date`",
                        "python ./artifacts/scripts/
build.py train"
                ]
            },
...
```

Alternatively, the `evaluation_build` project supplies the `evaluate` parameter to the `build.py` script to make an API call to SageMaker to run a processing Job that evaluates the trained ML model.

The final CodeBuild project we create is called `deployment_build`. Here, we define the runtime logic for the `deploy.py` file. You will recall from *Chapter 4, Continuous Integration and Continuous Delivery (CI/CD) for Machine Learning*, that the `deploy.py` script captures the execution parameters from the pipeline to deploy the SageMaker Endpoint Stack.

As you can see from the following code snippet, the `deployment_build` project synthesizes, or generates, the CloudFormation template for the Endpoint Stack, called `EndpointStack.template.json`.

After the template file has been created, the `deployment_build` project then executes the `deploy.py` script to generate the necessary CloudFormation parameters, required to deploy the stack template, and stores these parameters in the `params.json` file:

```
...
            "build": {
                "commands": [
                        "echo Synthesizing cdk
template",
                        "npx cdk synth -o output"
                ]
```

```
                },
                "post_build": {
                    "commands": [
                        "python ./artifacts/scripts/
deploy.py"
                    ]
                }
            },
            artifacts={
                "base-directory": "output",
                "files": [
                    "EndpointStack.template.json",
                    "params.json"
                ]
            }
...
```

Now that we have the relevant runtime logic to build the model artifact container image, preprocess the training data, train the ML model, and then evaluate the ML model's performance, we put can put these components together to construct the CI/CD pipeline to automate the process. Using the `pipeline` variable, we define the overall structure of the pipeline and, as you can see, the pipeline is comprised of four consecutive steps or **Pipeline Stages**:

1. Source
2. Build
3. Approval
4. Deploy

The **Source** stage refers to our CodeCommit repository, which is comprised of two branches. Updating any of these sources will trigger a release execution of the pipeline:

- The **main** branch contains the codified pipeline.
- The **model** branch contains the ML model artifacts.

The **Build** stage executes the four CodeBuild projects, as pipeline actions, to create the continuous integration phase of the process, and compiles or builds the ML model assets. The four pipeline actions are as follows:

1. The **ContainerBuild** action creates the container image, from the **model** source, and uploads it to the ECR repository.

2. The **Preprocess** action executes the `build.py` script to create a SageMaker processing job, whereby the raw training data is preprocessed to make it ready for model training.

3. The **Train** action also executes the `build.py` script, passing in the `train` parameter to create a SageMaker training job to build the optimized model.

4. Finally, the **Evaluate** action also calls the `build.py` script, passing in the `evaluate` parameter to create a SageMaker processing job that evaluates the performance of the trained model to assess its readiness for production.

The **Approval** stage will pause the pipeline by creating a manual decision gate, whereby the application owners will assess the model's performance results, and manually **Approve** or **Deny** the model for production. If the evaluation is denied, the pipeline execution halts and the cross-functional team assesses what source changes need to be made to improve the model. If the evaluation is approved, the pipeline automatically transitions to the **Deploy** stage.

The **Deploy** stage is essentially the continuous deployment phase of the process and, is comprised of pipeline actions:

1. The `DeploymentBuild` action is a CodeBuild project that synthesizes the endpoint CDK construct and executes the `deploy.py` script to gather the deployment parameters from the running pipeline.

2. The `DeployEndpoint` action deploys the synthesized CloudFormation template to create the endpoint stack and deploy the approved model into production.

Now that the various application components have been created, the final task is to configure the CDK application.

Creating the CDK application

The following steps will walk you through the process of adding the final code to complete the CDK application:

1. Using the navigation panel of the Cloud9 workspace, run the following commands to copy the `app.py` file from the companion GitHub repository to replace the template file created during the CDK initialization:

    ```
    $ cd ~/environment/abalone-cicd-pipeline/
    $ cp ~/environment/src/Chapter05/cdk/app.py .
    ```

2. Now, double-click on the `app.py` file so we can review it.

    As we review the `app.py` file, you can see that we import the necessary libraries, as well as the `EndpointStack()` and `PipelineStack()` classes, that we created earlier. Next, as you can see from the following code snippet, we define the parameters specific to our application, namely the name of the ML model, the name of the CodeCommit repository, and the current version of the AWS CDK we have installed:...

    ```
    MODEL = "abalone"
    CODECOMMIT_REPOSITORY = "abalone-cicd-pipeline"
    CDK_VERSION = "2.3.0"
    . . .
    ```

Next, we define the CDK application itself, using `cdk.App()`, and as the following code snippet shows, we declare an instance of the `EndpointStack()` class, while supplying the necessary CDK application parameters, including the current AWS region as well as the AWS account we are using:

```
. . .
EndpointStack(
    app,
    "EndpointStack",
    env=cdk.Environment(account=os.getenv("CDK_DEFAULT_
ACCOUNT"), region=os.getenv("CDK_DEFAULT_REGION")),
    model_name=MODEL
)
. . .
```

Lastly, as the next code snippet shows, we declare an instance of `PipelineStack()`, and also supply the required CDK application parameters required by the construct:

```
...
PipelineStack(
    app,
    CODECOMMIT_REPOSITORY,
    env=cdk.Environment(account=os.getenv("CDK_DEFAULT_
ACCOUNT"), region=os.getenv("CDK_DEFAULT_REGION")),
    model_name=MODEL,
    repo_name=CODECOMMIT_REPOSITORY,
    cdk_version=CDK_VERSION
)
...
```

We've now created our CDK application and it's ready to be deployed. The next section will show you how to deploy the application.

Deploying the pipeline application

Deploying the application on AWS is relatively straightforward. The following steps will walk you through this process, using the Cloud9 workspace terminal:

1. Before deploying the application, we need to finalize the application dependencies. Since the ML model will require raw training data, we need to download the data from the UCI repository. Run the following commands to get the training data:

```
$ cd ~/environment/abalone-cicd-pipeline/ && mkdir -p
artifacts/data
$ wget -c -P artifacts/data https://archive.ics.uci.edu/
ml/machine-learning-databases/abalone/abalone.data
```

2. Next, we need to configure the CDK environment by specifying the AWS region we are using, as well as our AWS account. Run the following command to bootstrap the CDK environment:

```
$ cdk bootstrap aws://${CDK_DEFAULT_ACCOUNT}/${CDK_
DEFAULT_REGION}
```

3. Now we can deploy the pipeline application, by running the following command:

```
$ cdk deploy abalone-cicd-pipeline
```

> **Note**
>
> When prompted **Do you wish to deploy these changes (y/n)?**, enter y, and
> hit *Enter*.

The application should take around 2 minutes to deploy and you can view the progress
within the Cloud9 terminal or the CloudFormation console (https://console.aws.
amazon.com/cloudformation/).

After the CloudFormation stack has been completed, we can trigger a pipeline release
by committing the application code into the CodeCommit repository. Run the following
commands to create an initial commit of the CDK application:

```
$ git add -A
$ git commit -m "Initial commit of Pipeline Artifacts"
$ git push
```

Now you can view the pipeline in the CodePipeline console (https://console.aws.
amazon.com/codesuite/codepipeline/) and click on **Pipeline** in the console's
navigation pane. *Figure 5.1* shows an example of what you might see:

Figure 5.1 – CodePipeline console

As you can see from *Figure 5.1*, **abalone-cicd-pipeline** has **Failed**. If you click on the pipeline to open the details, you will see that the **Source** stage of it has failed, as shown in *Figure 5.2*.

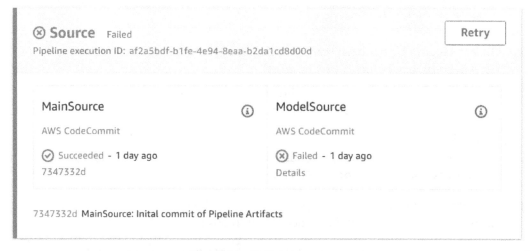

Figure 5.2 – Failed Source stage

Figure 5.2 shows the **Source** stage of the pipeline, and as you can see from the example, the **ModelSource** action has failed. This is because the ML practitioner team hasn't created any model source artifacts yet. In the next section, we will work through creating these artifacts.

Building the ML model artifacts

Up to this point, we have focused on the various tasks that are typically performed by the application development teams, creating a CDK application for the overall structure of the automated process. In this section, we will continue this undertaking, but from the perspective of the ML practitioners, whereby we will create the ML model itself, as well as the artifacts responsible for executing the data processing, ML model training, and ML model evaluation processes. The following steps will show you how an ML practitioner might do this:

1. Using your AWS account, open the SageMaker console (`https://console.aws.amazon.com/sagemaker/home`).

2. Using the left-hand menu panel, click on the **Studio** option to open the **SageMaker Domain** dashboard.

3. In the **SageMaker Domain** dashboard, click on the **Launch app** drop-down menu and select **Studio** to launch the Studio UI in the browser.

> **Note**
> You should have a SageMaker Domain already configured in the SageMaker console. If not, please refer to the *Getting started with SageMaker Studio* section in *Chapter 2, Automating Machine Learning Development Using SageMaker Autopilot.*

4. Once the Studio UI has been launched, click the **File** menu, then click **New**, and select **Terminal**, to launch a new terminal.

5. Next, we will clone the companion GitHub repository to access the pre-built artifacts. In the terminal, run the following commands:

```
$ cd ~
$ git clone https://github.com/PacktPublishing/Automated-
Machine-Learning-on-AWS src
```

6. Now, we will clone the pipeline repository. Run the following commands in the terminal, get the address of the CodeCommit repository, and clone it:

```
$ CLONE_URL=$(aws codecommit get-repository --repository-
name abalone-cicd-pipeline --query "repositoryMetadata.
cloneUrlHttp" --output text)
$ git clone $CLONE_URL
```

7. Run the following commands to create the model artifact branch:

```
$ cd ~/abalone-cicd-pipeline/
$ git checkout -b model
```

8. Before we can create the model artifacts, we need to clear out the existing code from the new branch. Run the following command to create a fresh branch:

```
$ git rm -rf .
```

9. Copy the pre-built Jupyter notebook from the cloned companion GitHub repository to the model branch by running the following command:

```
$ cp ~/src/Chapter05/Notebook/Abalone\ CICD\ Example.
ipynb .
```

10. In the navigation panel of the Studio UI, double-click on **Abalone CICD Example. ipynb** to open the notebook for review.

> **Note**
> The Jupyter Notebook requires the Python 3 (Data Science) kernel, which may take up to 2 minutes to launch.

11. Once the Jupyter kernel has started, from the **Kernel** menu, click the **Restart Kernel and Run all Cells …** option to execute all the notebook code cells.

Once all the code cells have been executed, you should see that we have followed a similar methodology to the one we used in *Chapter 3*, *Automating Complicated Model Development with AutoGluon*, where the ML practitioner built a deep learning container image for AutoGluon. In the same way, we have created the necessary component files to construct the container image that represents our model artifact. Hence, you should now see five new files in the left-hand navigation panel of the Studio UI:

- `model.py`
- `app.py`
- `nginx.conf`
- `wsgi.py`
- `Dockerfile`

Let's review what each of these component files does, within the context of our container image.

Reviewing the modeling file

The `model.py` file is primarily responsible for all tasks pertaining to the ML model itself. As you will see, there are three central Python functions to handle the tasks of preparing the training data, training the ML model, and evaluating the ML model. For example, the `preprocess()` function will take the raw data, preprocess the dataset by encoding the categorical values, and then split the data into a training (80% of the data) dataset, validation (15% of the data) dataset, and testing (5% of the data) dataset.

Once the data has been processed, we use the `train()` function to compile and fit the TensorFlow model to the data. The trained model is then saved for evaluation and inference.

The last function we will create as part of the model runtime is the `evaluate()` function. This function will load the model, using the `load_model()` function, evaluate the quality of the trained model, and then save the report by means of the `save_report()` function.

So, by creating these core functions in the modeling file, we have a single runtime script that handles the primary function of producing a production model candidate for approval.

Let's further review just how the modeling file is used, as well diving into app.py next.

Reviewing the application file

Within the context of continuous integration and continuous deployment, the model artifact will perform two functions, model training and model hosting. To wrap the runtime logic for determining which of these two tasks the model artifact performs, the second file generated by the notebook is the app.py file. This file serves as the main entry point to the container image and depending on the arguments passed to this entry point, the runtime logic within the app.py file will determine whether to train or host the model.

As an example, if you refer to the build.py file that we created in the previous section, and as shown in the following code snippet, to preprocess the data as a SageMaker processing job, the handle_data() function calls the create_processing_job() SageMaker API. As part of the AppSpecification parameter for the API call, we provide the container image URI, along with the preprocess argument for the app.py entry point:

```
...
def handle_data(model_name=None, execution_id=None):
    try:
        response = sagemaker_client.create_processing_job(
            ProcessingJobName=f"{model_name}-ProcessingJob-
{execution_id}",
            ProcessingResources={
                'ClusterConfig': {
                    'InstanceCount': 1,
                    'InstanceType': 'ml.m5.xlarge',
                    'VolumeSizeInGB': 30
                }
            },
            StoppingCondition={
                'MaxRuntimeInSeconds': 3600
            },
            AppSpecification={
```

```
            'ImageUri': f"{image_uri}:latest",
            'ContainerEntrypoint': ["python", "app.py",
"preprocess"]
        },
...
```

So, when SageMaker initializes the container image to run the processing job, it will run app.py as the entry point, and supply the preprocess argument. Now if we refer to the __main__ routine within the app.py file, as highlighted in the next code snippet, we can see that when the preprocess argument is provided, the main program will in turn execute the preprocess() function within the model.py file:

```
...
if __name__ == "__main__":
    print(f"Tensorflow Version: {tf.__version__}")
    if len(sys.argv) < 2 or ( not sys.argv[1] in [ "serve",
"train", "preprocess", "evaluate"] ):
        raise Exception("Invalid argument: you must specify
'train' for training mode, 'serve' for predicting mode,
'preprocess' for preprocessing mode or 'evaluate' for
evaluation mode.")
    preprocess = sys.argv[1] == "preprocess"
    train = sys.argv[1] == "train"
    evaluate = sys.argv[1] == "evaluate"
    if preprocess:
        model.preprocess()
    elif train:
        model.train()
    elif evaluate:
        model.evaluate()
    else:
        cpu_count = multiprocessing.cpu_count()
        model_server_timeout = os.environ.get('MODEL_SERVER_
TIMEOUT', 60)
        model_server_workers = int(os.environ.get('MODEL_
SERVER_WORKERS', cpu_count))
        start_server(model_server_timeout, model_server_
workers)
...
```

Now, if you continue to examine the previous code snippet, you can further see that the same overall concept applies for the model training, as well as the model evaluation processes, if the `train` or `evaluate` arguments are supplied to the entry point.

Alternatively, if none of these arguments are supplied, the application wrapper will perform the hosting task, consequently providing the trained model as a hosted endpoint. Next, we'll examine the additional artifact files necessary for hosting the model.

Reviewing the model serving files

Since the model will be served using Python's Flask framework (`https://flask.palletsprojects.com`), we need to add web serving components, such as NGINX (`https://www.nginx.com/`) and WSGI (`https://www.palletsprojects.com/p/werkzeug/`). The configurations for these web serving components are stored in the `nginx.conf` and `wsgi.py` files.

Reviewing the container build file

The final file we created within the notebook is the `Dockerfile`. This file will execute the container build instructions to download a TensorFlow 2.5 deep learning container from AWS, configure the web serving packages, and copy the model artifacts into the container.

So now that we've reviewed the files that make up the container image artifact, we can go ahead and update the source code repository.

Committing the ML artifacts

The final task that the ML practitioner performs is to commit the model artifacts into the CodeCommit repository and thus trigger a release of the CI/CD pipeline. To do this, run the following commands using the **Terminal** tab of the Studio UI:

```
$ cd ~/ abalone-cicd-pipeline/
$ git add -A
$ git config --global user.email "<ENTER YOUR EMAIL ADDRESS>"
$ git config --global user.name "<ENTER YOUR NAME>"
$ git commit -m "Initial commit of model artifacts"
$ git push --set-upstream origin model
```

> **Note**
>
> Make sure to substitute your unique email address and username when committing the new artifacts. This way, any code changes that trigger a new release can be tracked.

If we assume that the ML practitioner has executed the appropriate unit tests, to ensure that the data processing, model training, and evaluation functions work, and since we have committed these artifacts into the CodeCommit repository, we can finally automate the deployment of the ML model into production. The next section will review the model release process.

Executing the automated ML model deployment

Reviewing the pipeline execution is done through the CodePipeline console (`https://console.aws.amazon.com/codesuite/codepipeline/home`), and then by clicking on the **abalone-cicd-pipeline** name. Once the pipeline dashboard opens, the first thing you will immediately see (as shown in *Figure 5.3*) is that the **Source** stage has succeeded.

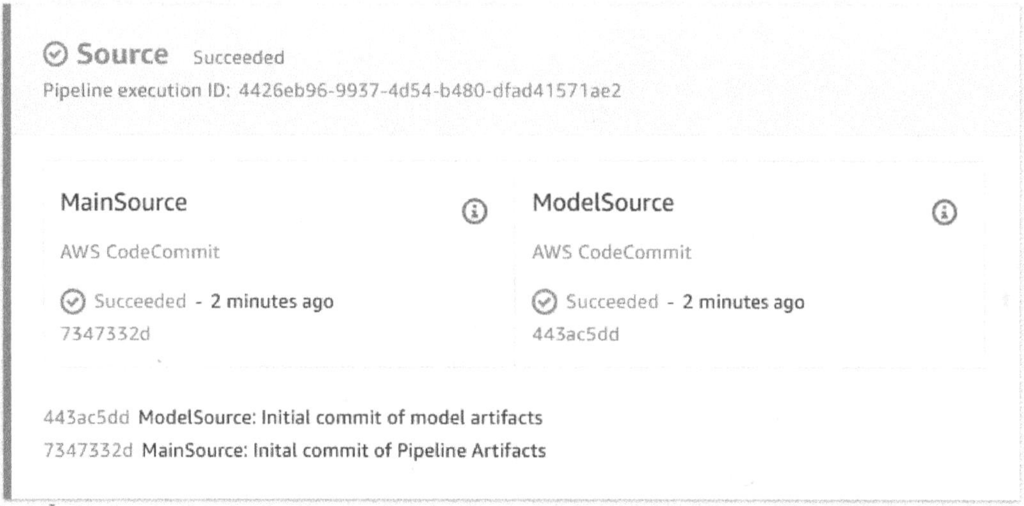

Figure 5.3 – Succeeded Source stage

Once the **Source** stage succeeds, the pipeline automatically moves onto the **Build** stage to essentially execute the continuous integration phase of the CI/CD process. *Figure 5.4* shows the four stage actions that cover continuous integration.

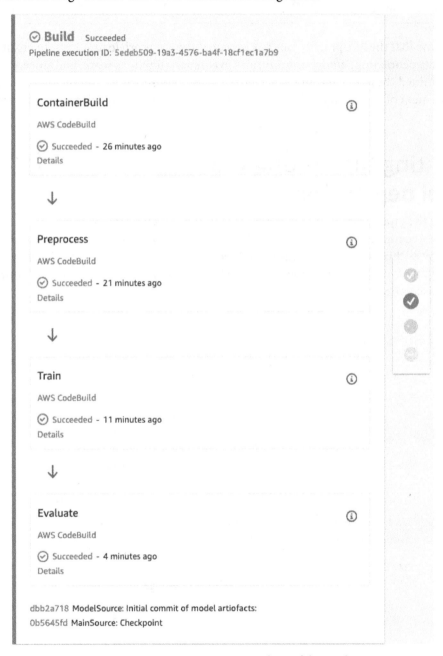

Figure 5.4 – Continuous integration phase of the pipeline

From *Figure 5.4*, there are a few items to take note of. Firstly, note the **Pipeline execution ID**. This ID is used to track the lineage of the ML model's release, as it is embedded into each the name of the various SageMaker jobs. The ID is also used as an S3 folder containing all the relevant assets used by the pipeline. For example, by opening the ID folder in the S3 console (`https://s3.console.aws.amazon.com`), you can see the model assets, training, validation, and testing datasets, as well as the model evaluation report.

Secondly, by clicking on the **Details** link for every stage action, you can review the output from each CodeBuild project. Recall that a CodeBuild project was created to execute the container image build, data processing, model training, and model evaluation steps of the pipeline.

So, if you click on the **Details** link for the **Train** stage action, you are redirected to the **Build** status dashboard for the `ModelTrainingBuild` project. If you scroll down, you'll see the output from the **Build logs**. *Figure 5.5* shows an example of the CodeBuild log output for the ML model training.

```
94
95  [Container] 2021/11/24 15:27:09 Running command python ./artifacts/scripts/build.py train
96  INFO: [build.py:251] Executing TRAIN task
97  INFO: [build.py:243] Task: train, Status: InProgress
98  INFO: [build.py:243] Task: train, Status: InProgress
99  INFO: [build.py:243] Task: train, Status: InProgress
100 INFO: [build.py:243] Task: train, Status: InProgress
101 INFO: [build.py:243] Task: train, Status: InProgress
102 INFO: [build.py:243] Task: train, Status: InProgress
103 INFO: [build.py:243] Task: train, Status: InProgress
104 INFO: [build.py:243] Task: train, Status: InProgress
105 INFO: [build.py:266] Task: train, Final Status: Completed
106
```

Figure 5.5 – CodeBuild log output for model training

As you can see from *Figure 5.5*, the CodeBuild project executes the `build.py` script and supplies the `train` argument. You will recall that the `train` argument instructs the `build.py` file to execute the `handle_train()` function, whereby SageMaker is instructed to run a training Job, and use the model artifact container image to train the ML model.

> **Note**
> Since the various CodeBuild projects execute API calls to trigger SageMaker jobs for data processing, model training, and model evaluation, you can review the SageMaker specific logs using the SageMaker console (`https://console.aws.amazon.com/sagemaker/home`).

Lastly, by clicking on the source code commit IDs of both the **MainSource** and **ModelSource** branches of the pipeline, you can track what code changes were made for the release as well as who made those code changes.

So, once the continuous integration phase of the pipeline has been completed and the trained model evaluated, the pipeline pauses to wait for manual approval. *Figure 5.6* shows an example of the **Approval** stage:

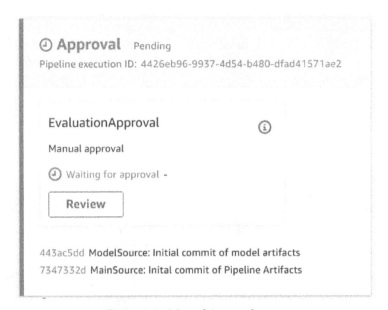

Figure 5.6 – Manual Approval stage

As you see from *Figure 5.6*, the pipeline is in a **Pending** state, waiting for the use case's acceptance criteria to be met, in order to proceed with the model's deployment. It is at this stage that the various application owners review the quality of the trained model candidate and determine whether or not the model is considered to be production-grade.

Since the evaluation report is a pipeline asset, it can be viewed using the S3 console. By opening the pipeline's S3 bucket, expanding the folder for the pipeline's execution ID, and then opening the evaluation sub-folder, the application owners can then open the evaluation.json file to review the evaluation report. The following is an example of what evaluation.json might look like:

```
{
    "regression_metrics": {
        "rmse": {
            "value": 1.4838999769750487
```

```
        },
        "mse": {
            "value": 2.20195914166655
        }
    }
}
```

Within the report, the application owners can see the rmse and mse evaluation metric results to decide if they approve or reject the model. This determination is then applied to the pipeline, by clicking the **Review** button, within the CodePipeline console, and adding any optional comments. The application owners can then click either the **Reject** or **Approve** buttons. *Figure 5.7* shows an example of what the pipeline **Review** process might resemble:

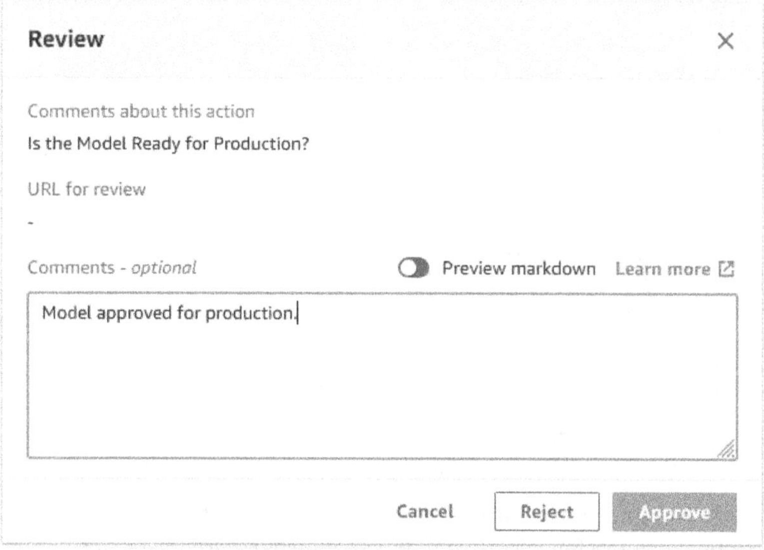

Figure 5.7 – Pipeline Review

Once the pipeline review has been approved, the pipeline execution proceeds onto the continuous deployment phase. It is at this point that the appropriate CloudFormation parameters are captured from the current pipeline execution, by the **DeploymentBuild** action, and the SageMaker endpoint is deployed using the **DeployEndpoint** action. *Figure 5.8* shows an example of the continuous deployment phase of the pipeline:

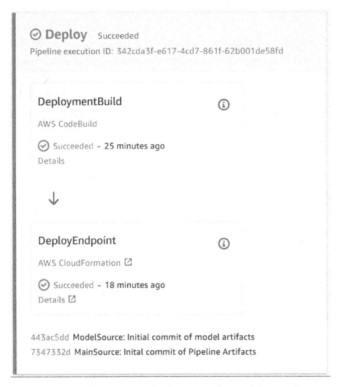

Figure 5.8 – Continuous deployment phase of the pipeline

As shown in *Figure 5.8*, once the **DeployEndpoint** action is complete, we now have an approved production model that can be integrated into the *Age Calculator* application, to serve abalone age predictions.

Since the CDK application artifacts and the ML artifacts exist in their own dedicated branch of the code repository, any further development on the pipeline or ML artifacts is owned and managed by the respective application team or ML practitioners. Any fixes or updates made to these branches from the feedback loop will cause a release change to the CI/CD pipeline and deploy a new version of the ML model into production.

So, now that we have shown how to continuously integrate, and continuously deploy the *Age Calculator* example, we can clean up the various resources.

Cleanup

To save on AWS resource costs, it is recommended that the deployment and pipeline assets are deleted. The following steps will guide you through this process:

1. Open the CloudFormation console (`https://console.aws.amazon.com/cloudformation/home`) and click on **EndpointStack** to open the stack details.

2. Now click on the **Delete** button to delete the SageMaker endpoint, endpoint configuration, and the trained model artifact.

3. Once the stack has been deleted, repeat the same process for the **abalone-cicd-pipeline** stack.

> **Note**
>
> Since the pipeline's S3 bucket and the abalone ECR repository are not empty, the stack deletion should fail. You will have to manually empty these resources and then try to delete the stack. You may also delete the **abalone-cicd-pipeline** CodeCommit repository. However, do not delete the Cloud9 environment as we will be using this in the next chapter.

Summary

In this chapter, we continued to build on the CDK application we started in *Chapter 4, Continuous Integration and Continuous Delivery (CI/CD) for Machine Learning*. In doing so, you were further presented with how to deploy the CDK application and automate the deployment of an optimized ML model.

You were also introduced to the importance of an agile, cross-function team as being integral to the success of an automated ML solution. We saw how these various teams bridged the gap between the ML modeling process (from the perspective of ML practitioners), all the way to automated model deployment (from the perspective of application development and operations teams).

Additionally, in this chapter, you saw how the AWS development tools, namely CodeCommit, CodeBuild, and CodePipeline, can be used to orchestrate the CI/CD process. Though the hands-on example, you saw for yourself how the typical ML process introduced in *Chapter 1, Getting Started with Automated Machine Learning on AWS*, can be integrated into the DevOps methodology, using the CI/CD process to create a foundation for MLOps.

In the next few chapters, we will continue to expand on the concepts of **Processes**, **Tools**, and **People** to build up to an automated machine learning software development lifecyle for the *Age Calculator* use case.

Section 3: Optimizing a Source Code-Centric Approach to Automated Machine Learning

This section will introduce you to the limitations of the overall CI/CD process and how to further integrate the role of the ML practitioner into the pipeline build process. The section will also introduce how this role integration streamlines the automation process and present you with an optimized methodology by introducing you to AWS Step Functions.

This section comprises the following chapters:

6
Automating the Machine Learning Process Using AWS Step Functions

In the first three chapters of the book, we saw a fundamental process flaw that can impact the automation of an ML use case, namely the handover of a production-grade model, produced by the ML practitioner, to the application development and operations teams. In *Chapter 4, Continuous Integration and Continuous Delivery (CI/CD) for Machine Learning*, we examined how this issue could be addressed by combining the ML processing into the DevOps process using the CI/CD process methodology.

While that solution inherently addresses the issue, you can also conclude that in terms of overall ownership, the development or platform teams were primarily responsible for building the majority of the final solution. For instance, you will recall from the CI/CD example we used in the previous chapter, the application development teams built the pipeline foundation, as well as the integrations for offloading the data processing, model training, and model evaluation tasks to SageMaker. These integrations require the development teams to have a fundamental understanding of ML in general, and the overall ML process.

While the example did show the ML practitioner providing more than just an optimized model to this cross-functional team in the form of a packaged container image, the fact remains that most of the solution's development was still the responsibility of the application and platform developers, thus requiring them to, in essence, be ML practitioners themselves.

Granted, the disproportion of assigned responsibilities may be because not all ML practitioners are themselves DevOps engineers and, organizationally speaking, not all ML practitioners are part of the same team as the developers and infrastructure staff.

Either way, *how do we further streamline the ML automation process without having to further skill up the development and ML teams, or change the organizational structure?*

Answering this question will be the primary focus of this chapter, where we will continue to build upon the foundation we established in the previous chapter and continue to streamline the *Age Calculator* example. To this end, we will cover the following topics:

- Introducing AWS Step Functions

- Using the AWS Step Functions Data Science SDK for CI/CD

- Building the CI/CD pipeline resources

Technical requirements

Here is a list of the technical requirements for this chapter:

- A web browser (for the best experience, it is recommended that you use Chrome or Firefox).

- Access to the AWS account that you used in *Chapter 4, Continuous Integration and Continuous Delivery (CI/CD) for Machine Learning*.

- Access to the Cloud9 development environment we used in *Chapter 4, Continuous Integration and Continuous Delivery (CI/CD) for Machine Learning*.

- We will once again be working within the usage limits of the AWS Free Tier to avoid exceeding unnecessary costs.

- Source code examples and Jupyter Notebooks are provided in the companion GitHub repository for this chapter (`https://github.com/PacktPublishing/Automated-Machine-Learning-on-AWS/tree/main/Chapter06`). The code examples should already be available in the Cloud9 development environment. If not, refer to the section entitled *Developing the Application Artifacts* in *Chapter 4, Continuous Integration and Continuous Delivery (CI/CD) for Machine Learning*.

Introducing AWS Step Functions

At re:Invent 2016, AWS announced the **Step Functions** service as a way to orchestrate common business processes by creating a workflow. A workflow, also referred to as a **state machine**, is essentially a series of event-driven steps, or **States**, that denote a single process unit. By chaining these units of work together we are effectively creating an automated process to accomplish an overall goal.

In the case of automating the ML process, we can create a state machine that chains together individual steps to process the training data, train an ML model, evaluate the trained model's performance, and even deploy the model into production.

The advantage of using Step Functions for the ML process, or automating any workflow for that matter, is that we can re-direct the flow based on conditions our outcomes of each step. For example, if a specific step within the workflow fails, we can retry it or redirect the overall flow to follow some alternate process logic.

Creating a state machine

To create the overall workflow, we start by creating the individual states within the state machine. This is achieved by defining these states using the Amazon States Language (`https://states-language.net/spec.html`). The States Language is a JSON-based schema whereby you manually define each state as a JSON object. The following code shows an example of what a state machine might look like when using the States Language to define it:

```
{
    "Comment": "A simple minimal example of the States
language",
    "StartAt": "Hello World",
    "States": {
    "Hello World": {
      "Type": "Task",
      "Resource": "arn:aws:lambda:us-east-1:123456789012:functi
on:HelloWorld",
      "End": true
    }
  }
}
```

> **Note**
>
> This example is provided under the Apache License, Version 2.0, and is derived directly from the online copy of the Amazon States Language specification (`https://states-language.net/spec.html`).

Figure 6.1 shows the graphical representation of the workflow, or state machine definition, derived from the States Language JSON schema:

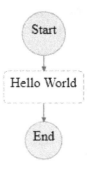

Figure 6.1 – State machine definition

As you can see from *Figure 6.1*, as well as the JSON schema example, we've defined a state called Hello World. We've further specified that this type of state is a Task state, whereby its unit of work is to execute an AWS Lambda function. Additionally, we've defined the workflow to start with this task and end after the task has been executed.

So, while this may seem very straightforward, as we will see in *Chapter 7, Building the ML Workflow Using AWS Step Functions*, when we start defining states that include retries and failures and incorporate different choice paths, the resultant JSON schema definition can be extremely intricate.

Addressing state machine complexity

Since creating the Step Function service, AWS has provided a couple of mechanisms to overcome the complexities associated with manually defining a state machine using the States Language.

For example, in July of 2019, AWS introduced the **AWS Toolkit for Visual Studio Code** (`https://aws.amazon.com/visualstudiocode/`). As part of this toolkit, AWS provided developers with the ability to define, visualize, execute, and update state machines from within the VS Code IDE. Along with code completion and state machine validation, developers can overcome some of the complexities associated with defining state machines with the States Language when using VS Code.

Additionally, in March of 2021, AWS introduced the ability to define state machines using the **YAML Ain't Markup Language** (**YAML**) serialization language instead of JSON, thus making it easier for the developer to build state machines if YAML is their serialization language of choice.

Furthermore, in July 2021, AWS announced **Workflow Studio**. This is a visual workflow design tool that allows developers to use a graphical design tool, within the AWS console, to build state machines by simply dragging and dropping workflow and task states onto a canvas, and integrate them using a minimal amount of code, consequently making it easier for developers to build complicated workflows.

However, even though these added capabilities make it easier to define ML-based workflows, the question remains: *who is ultimately is responsible for defining the state machine schema? Is it the application development teams or the ML practitioner?*

In the next section, we will evaluate using Step Functions capabilities to help the data scientist and ML practitioner to further streamline and automate the ML workflow.

Using the Step Functions Data Science SDK for CI/CD

In November 2019, AWS introduced the **AWS Step Functions Data Science SDK for Amazon SageMaker**. This SDK allows data science and ML practitioners to programmatically construct Step Function workflows to deliver production-grade ML models. The SDK is designed to be used within a Jupyter Notebook to construct a process that delivers a reproducible ML experiment in the form of a Step Functions workflow, as opposed to reproducing the experiment itself.

Basically, what this means is instead of the ML practitioner exploring data, building algorithms, training models, and evaluating the trained model's performance, they instead construct a state machine to accomplish these tasks automatically. On top of this, the resulting state machine is constructed programmatically, instead of manually defining it with the States Language specification. Therefore, to answer the questions raised in the previous section, the ML practitioner can now own the task of defining the automation process to produce a production-grade ML model.

How so?

Previously we saw how the ML practitioner delivered pre-packaged container images as an artifact to the CI/CD process. This artifact contained the various runtime processes to handle the data, train the model, and evaluate the model's performance. We also saw how the development and operations teams had to re-factor the CI/CD pipeline to incorporate the typical ML process.

Now, in the spirit of a cross-functional team, the ML practitioner can rather deliver a state machine that automates the entire process as a CI/CD pipeline artifact. Plus, by using the Data Science SDK, the state machine can be programmatically defined without having to up-skill the ML practitioner team. On the other hand, the development teams don't have to up-skill their ML knowledge to incorporate the ML process into the CI/CD pipeline.

To demonstrate exactly how an agile and cross-functional team would create this solution, let's re-factor the previously used CI/CD process from scratch. *Figure 6.2* provides an overview of the resulting re-factored process.

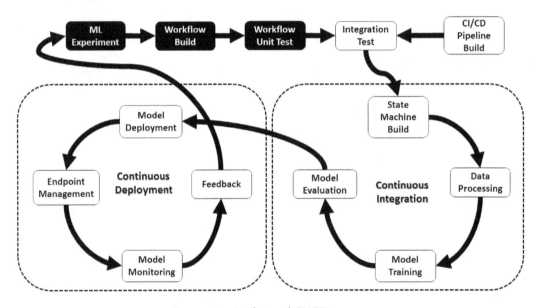

Figure 6.2 – Re-factored CI/CD process

As you can see in *Figure 6.2*, the process is slightly different when you compare it to the process outlined in *Figure 4.5* in *Chapter 4, Continuous Integration and Continuous Delivery (CI/CD) for Machine Learning*. In the re-factored process, the ML practitioner takes on more of a significant role. While the CI and CD phases remain mostly the same, the ML practitioner is now responsible for developing the workflow assets that will orchestrate these processes.

For example, in the re-factored process, the ML practitioner is not only responsible for providing the optimal model artifacts from the ML experiment, but now they are also responsible for building and testing the automated ML process using the Step Functions Data Science SDK. This automated ML workflow artifact is responsible for preprocessing the training data, training the ML model, and evaluating whether or not the model is ready for production.

However, before the workflow assets can be integrated, the development engineers must also build the CI/CD pipeline. The next section will walk you through how to do that.

Building the CI/CD pipeline resources

To begin re-factoring the *Age Calculator* use case, we are going to work through the initial setup steps from the perspective of the development and operations teams. We will be using the same Cloud9 development environment that we created in *Chapter 4, Continuous Integration and Continuous Delivery (CI/CD) for Machine Learning*, to perform the following tasks:

- Updating the development environment
- Creating the pipeline artifact repository
- Building the pipeline application artifacts
- Deploying the CI/CD pipeline

Let's get started.

Updating the development environment

Start by logging into the same AWS account you've been using up to this point and open the AWS Cloud9 console (`https://console.aws.amazon.com/cloud9`). Under **Your environments**, click the **Open IDE** button to launch the **MLOps-IDE** development environment.

> **Note**
>
> If you have not provisioned the **MLOps-IDE** environment, please refer to the *Preparing the development environment* section of *Chapter 4, Continuous Integration and Continuous Delivery (CI/CD) for Machine Learning*.

To update the environment, execute the following steps:

1. Run the following command to ensure that we have version 2.3.0 of the CDK installed:

    ```
    $ cdk --version
    ```

    > **Note**
    >
    > At the time of writing, the latest version of the CDK is **2.3.0 (build beaa5b2)**. If you are not running this version within the Cloud9 environment, refer to *Chapter 4, Continuous Integration and Continuous Delivery (CI/CD) for Machine Learning*, for instructions on how to install it.

Now that we have updated the environment to the latest version of the CDK, we can create the source code repository.

Creating the pipeline artifact repository

Execute the following steps to create a new `abalone-cicd-pipeline` CodeCommit repository:

1. Using the workspace terminal, run the following CLI command to ensure that the CLI region settings are correct. Make sure to replace `<REGION>` with the AWS Region you are currently using:

    ```
    $ aws configure set region <REGION>
    ```

    > **Note**
    >
    > Since we are re-factoring the previous solution, we will be using the same repository name as in *Chapter 4, Continuous Integration and Continuous Delivery (CI/CD) for Machine Learning*. Therefore, make sure that you have cleaned up any existing resources. If not, make sure to manually delete the `abalone-cicd-pipeline` CodeCommit repository in the CodeCommit management console (`https://console.aws.amazon.com/codesuite/codecommit/repositories`).

2. Create the new CodeCommit repository called `abalone-cicd-pipeline`, using the following command:

```
$ aws codecommit create-repository --repository-name
abalone-cicd-pipeline --repository-description "Automated
ML on AWS using the Step Functions Data Science SDK"
```

3. Next, capture the URL for the newly created repository in order to clone it. Run the following command to create the CLONE_URL parameter:

```
$ CLONE_URL=$(aws codecommit get-repository --repository-
name abalone-cicd-pipeline --query "repositoryMetadata.
cloneUrlHttp" --output text)
```

4. Run the following command to clone the empty repository, locally:

```
$ git clone $CLONE_URL
```

Now that we have our new project repository, we can proceed to the next task of building out the application artifacts.

Building the pipeline application artifacts

Use the following steps to build out the pipeline application:

1. Initialize a new CDK project by running the following command:

```
$ cd ~/environment/abalone-cicd-pipeline && cdk init
--language python
```

2. Set the primary branch of the source repository by running the following commands:

```
$ git add -A
$ git commit -m "Started Pipeline Project"
$ git branch main
$ git checkout main
```

3. Configure the Python environment by running the following commands:

```
$ source .venv/bin/activate
$ python -m pip install -U pip pylint boto3
$ pip install -r requirements.txt
```

You will recall from the original example in *Chapter 4*, *Continuous Integration and Continuous Delivery (CI/CD) for Machine Learning*, that we started building out the CDK application by looking at the final goal for the application and working backward to develop the artifacts that accomplish the objective. Since the final goal of the pipeline is to have a production-grade ML model, hosted as a SageMaker endpoint, we need to build out what the endpoint stack looks like. By working with the ML practitioner team, we (as the development team) can gather the functional requirements to build out the endpoint using the following steps:

1. Using the navigation panel of the Cloud9 workspace, expand the `abalone-ci-cd-pipeline` folder, then right-click on the `abalone_cicd_pipeline` folder and select the **New File** option.

2. Name the newly created file `abalone_endpoint_stack.py` and double-click on it for editing.

> **Note**
>
> A complete copy of the `abalone_endpoint_stack.py` is available for review in the companion GitHub repository (`https://github.com/PacktPublishing/Automated-Machine-Learning-on-AWS/tree/main/Chapter06/cdk`). The file should also available be available for review in the `~/environment/src/Chapter06/cdk/` folder within the Cloud9 environment.

3. Inside the Python file, add the following code to import the CDK modules for the endpoint:

```
. . .
import aws_cdk as cdk
import aws_cdk.aws_sagemaker as sagemaker
. . .
```

4. Next, create a Python class called `EndpointStack()` as a CDK stack construct by adding the following code:

```
. . .
class EndpointStack(cdk.Stack):
    def __init__(self, app: cdk.App, id: str, *, model_
name: str=None, repo_name: str=None, **kwargs) -> None:
        super().__init__(app, id, **kwargs)
. . .
```

5. Now, we use the following code to create the required pipeline parameters for the S3 bucket and the pipeline execution ID. These parameters will be derived during the pipeline execution:

```
...
        bucket_name = cdk.CfnParameter(
            self,
            "BucketName",
            type="String"
        )

        execution_id = cdk.CfnParameter(
            self,
            "ExecutionId",
            type="String"
        )
...
```

6. Since we have instantiated the parameters for the endpoint construct, we can now define the endpoint configuration in the following code snippet. This configuration details what compute resources to run the endpoint on, as well as the trained model to host for the endpoint, plus where to store the inference and response data for model monitoring. The following code shows how to instantiate the endpoint_config variable as a CfnEndpointConfig():

```
...
        endpoint_config = sagemaker.CfnEndpointConfig(
            self,
            "EndpointConfig",
            endpoint_config_name="{}-config-{}".
format(model_name.capitalize(), execution_id.value_as_
string),
            production_variants=[
                sagemaker.CfnEndpointConfig.
ProductionVariantProperty(
                    initial_instance_count=2,
                    initial_variant_weight=1.0,
                    instance_type="ml.m5.large",
                    model_name="{}-{}".format(model_name,
```

```
execution_id.value_as_string),
                        variant_name="AllTraffic"
            )
        ],
...
```

7. Continuing from the previous code snippet, we continue defining the CfnEndpointConfig() and specify the data_capture_config parameter. As the following code snippet shows, we specify a DataCaptureConfigProperty() that configures the endpoint to capture 100% of the input data to the endpoint, as well as the output data from the endpoint, to S3:

```
...
            data_capture_config=sagemaker.
CfnEndpointConfig.DataCaptureConfigProperty(
                capture_content_type_header=sagemaker.
CfnEndpointConfig.CaptureContentTypeHeaderProperty(
                    csv_content_types=[
                        "text/csv"
                    ]
                ),
                capture_options=[
                    sagemaker.CfnEndpointConfig.
CaptureOptionProperty(capture_mode="Input"),
                    sagemaker.CfnEndpointConfig.
CaptureOptionProperty(capture_mode="Output")
                ],
                destination_s3_uri="s3://{}/endpoint-
data-capture".format(bucket_name.value_as_string),
                enable_capture=True,
                initial_sampling_percentage=100.0
            )
        )
...
```

8. The final part of the construct is the endpoint itself. We use the following code to declare the `endpoint`, and then save the file:

```
...
        endpoint = sagemaker.CfnEndpoint(
            self,
            "AbaloneEndpoint",
            endpoint_config_name=endpoint_config.attr_
endpoint_config_name,
            endpoint_name="{}-Endpoint".format(model_
name.capitalize())
        )
        endpoint.add_depends_on(endpoint_config)
...
```

Since the endpoint deployment construct has been created, we now need to create the build script that captures execution parameters from a running CI/CD pipeline. The following steps will walk you through the process:

1. Using the Cloud9 terminal, run the following commands to create the necessary artifacts folder:

```
$ mkdir -p ~/environment/abalone-cicd-pipeline/artifacts/
scripts
```

2. Using the navigation panel, right-click on newly created `scripts` folder and select the **New File** option.

3. Name the file `deploy.py` and double-click on it for editing.

> **Note**
>
> A complete copy of the `deploy.py` is available for review in the companion GitHub repository (`https://github.com/PacktPublishing/Automated-Machine-Learning-on-AWS/tree/main/Chapter06/scripts`). The file should also available be available for review in the `~/environment/src/Chapter06/scripts/` folder within the Cloud9 environment.

4. In the `deploy.py` file, the first thing we do is add the following code snippet to import the Python libraries we'll be using:

```
. . .
import boto3
import logging
import os
import json
import sys
from botocore.exceptions import ClientError
. . .
```

5. Next, we add the following snippet of code to specify the global parameters, such as logging and the AWS Python SDK (`boto3`) clients for CodePipeline and SageMaker:

```
. . .
logger = logging.getLogger()
logging_format = "%(levelname)s: [%(filename)s:%(lineno)
s] %(message)s"
logging.basicConfig(format=logging_format, level=os.
environ.get("LOGLEVEL", "INFO").upper())
codepipeline_client = boto3.client("codepipeline")
sagemaker_client = boto3.client("sagemaker")
pipeline_name = os.environ["PIPELINE_NAME"]
model_name = os.environ["MODEL_NAME"]
. . .
```

6. The following code snippet shows a function called `get_execution_id()`. This function makes a call to the running CI/CD pipeline and returns the current execution ID. This ID is used to version the model that will be hosted as a SageMaker endpoint:

```
. . .
def get_execution_id(name=None, task=None):
    try:
        response = codepipeline_client.get_pipeline_
state(name=name)
        for stage in response["stageStates"]:
            if stage["stageName"] == "Deploy":
```

```
                for action in stage["actionStates"]:
                    if action["actionName"] == task:
                        return stage["latestExecution"]
    ["pipelineExecutionId"]
        except ClientError as e:
            error = e.response["Error"]["Message"]
            logger.error(error)
            raise Exception(error)
    ...
```

7. Lastly, the following code snippet instantiates the main program. This program creates and stores the execution ID and S3 bucket name in a params.json file. This JSON file will be used as input parameters to the endpoint deployment stack:

```
    ...
if __name__ == "__main__":
    task = "DeploymentBuild"
    execution_id = get_execution_id(name=pipeline_name,
    task=task)
    logger.info("Creating Stack Parameters")
    params = {
        "ExecutionId": execution_id,
        "BucketName": os.environ["BUCKET_NAME"]
    }
    try:
        with open(os.path.join(os.environ["CODEBUILD_SRC_
    DIR"], "output/params.json"), "w") as f:
            json.dump(params, f)
        logger.info(json.dumps(params, indent=4)),
        sys.exit(0)
    except Exception as error:
        logger.error(error)
        sys.exit(255)
    ...
```

8. After entering the previous code in the deploy.py file, make sure to save it.

At this point, we have created the necessary CDK constructs and supporting deployment code for hosting the model as a SageMaker endpoint. Now we can move on to creating the CI/CD pipeline construct by using the following steps:

1. Within the Cloud9 navigation panel, expand the `abalone_cicd_pipeline` folder and double-click on the `abalone_cicd_pipeline_stack.py` file for editing.

> **Note**
>
> A complete copy of the `abalone_cicd_pipeline_stack.py` file is available for review in the companion GitHub repository (`https://github.com/PacktPublishing/Automated-Machine-Learning-on-AWS/tree/main/Chapter06/scripts`). The file should also available be available for copying into the `abalone_cicd_pipeline` folder, from the `~/environment/src/Chapter06/cdk/` folder within the Cloud9 environment.

2. If you choose to create the file from scratch, delete any existing template code within the file so we can start with a blank file, and add the following code to import the necessary CDK modules for the pipeline construct:

```
...
import os
import aws_cdk.core as cdk
import aws_cdk.aws_codecommit as codecommit
import aws_cdk.aws_codepipeline as codepipeline
import aws_cdk.aws_codepipeline_actions as pipeline_
actions
import aws_cdk.aws_codebuild as codebuild
import aws_cdk.aws_iam as iam
import aws_cdk.aws_ecr as ecr
import aws_cdk.aws_s3 as s3
import aws_cdk.aws_s3_deployment as s3_deployment
import aws_cdk.aws_ssm as ssm
from constructs import Construct
...
```

3. Next, we add the following code snippet to create a new Python class called `PipelineStack()` as a CDK stack construct:

```
...
class PipelineStack(cdk.Stack):
    def __init__(self, scope: Construct, id: str, *,
model_name: str=None, repo_name: str=None, cdk_version:
str=None, **kwargs) -> None:
        super().__init__(scope, id, **kwargs)
...
```

4. The first component we define for the CDK construct is a placeholder for the CodeCommit repository. As you can see in the following code snippet, we reference the previously created `abalone-cicd-pipeline` CodeCommit repository using the `from_repository_name` method, and reference our repository variable as `repo_name`:

```
...
        code_repo = codecommit.Repository.from_
repository_name(
            self,
            "PipelineSourceRepo",
            repository_name=repo_name
        )
...
```

5. The next component to create is an IAM policy document. This policy will be used by the IAM role to not only execute the state machine but also access the various AWS resources used within the workflow. We define a variable called `workflow_policy_document` and create a Python dictionary to store the various IAM policy statements. The following code snippet shows an excerpt from the `Action` statement of the IAM policy. Here, you can see that we give the IAM role access to get the current CI/CD pipeline execution and invoke any Lambda functions contained within the state machine itself. We also give the role the ability to manage the state machine by providing it with the ability to create, delete, describe, and start any state machines:

```
...
            "Statement": [
                {
                    "Effect": "Allow",
```

```
                    "Action": [
                        "codepipeline:GetPipelineState",
                        "lambda:InvokeFunction",
                        "lambda:UpdateFunctionCode",
                        "lambda:CreateFunction",
                        "states:CreateStateMachine",
                        "states:UpdateStateMachine",
                        "states:DeleteStateMachine",
                        "states:DescribeStateMachine",
                        "states:StartExecution"
                    ],
                    "Resource": "*"
                },
...
```

> **Note**
>
> As you can see from the previous code snippet, the IAM policy is fairly *open* to the type of AWS resources used, since we specify "*" for all AWS resources. This is not recommended for a production use case. For more information on granting least privilege access to AWS resources, see the IAM security best practices documentation (`https://docs.aws.amazon.com/IAM/latest/UserGuide/best-practices.html#grant-least-privilege`).

6. Now that we've defined the policy statement, we can create the workflow execution role that uses the policy by declaring the `workflow_role` variable:

```
...
        workflow_role = iam.Role(
            self,
            "WorkflowExecutionRole",
            assumed_by=iam.CompositePrincipal(
                iam.ServicePrincipal("codebuild.
amazonaws.com")
            )
        )
...
```

7. The following code snippet shows how we apply the policy document to the `workflow_role` as an inline policy using the `attach_inline_policy()` method:

```
...
        workflow_role.attach_inline_policy(
            iam.Policy(
                self,
                "WorkflowRoleInlinePolicy",
                document=iam.PolicyDocument.from_
    json(workflow_policy_document)
            )
        )
...
```

8. Since any Lambda function within the state machine, as well as the state machine itself, requires an AWS service role, we add these two service principals to the `workflow_role`, giving the role access to assume the service roles. The following code snippet shows how we use the `add_statements()` method to provide the `AssumeRole` capability:

```
...
        workflow_role.assume_role_policy.add_statements(
            iam.PolicyStatement(
                actions=[
                    "sts:AssumeRole"
                ],
                effect=iam.Effect.ALLOW,
                principals=[
                    iam.ServicePrincipal("lambda.
    amazonaws.com"),
                    iam.ServicePrincipal("sagemaker.
    amazonaws.com"),
                    iam.ServicePrincipal("states.
    amazonaws.com")
                ]
            )
        )
...
```

9. Since the workflow will be executing various SageMaker functions to train, evaluate, and host the ML model, the following code snippet shows how we supply the `AmazonSageMakerFullAccess` managed IAM policy to the `workflow_role` by using the `add_managed_policy()` method:

```
...
        workflow_role.add_managed_policy(
            iam.ManagedPolicy.from_aws_managed_policy_
    name("AmazonSageMakerFullAccess")
        )
...
```

10. Now we add this **Amazon Resource Name** (**ARN**) as a parameter. This parameter will be used by the ML practitioner when defining the workflow, and it will be stored in the **AWS Systems Manager Parameter Store** (**SSM**). This way, when the ML practitioner teams need to reference the role, they can do so by using an API call to the parameter store:

```
...
        workflow_role_param = ssm.StringParameter(
            self,
            "WorkflowRoleParameter",
            description="Step Functions Workflow
    Execution Role ARN",
            parameter_name="WorkflowRoleParameter",
            string_value=workflow_role.role_arn
        )
        workflow_role_param.grant_read(workflow_role)
...
```

11. Next, we define a SageMaker execution role. SageMaker will use this role to access the various services it needs to process training data, train the model, and evaluate the model:

```
...
        sagemaker_role = iam.Role(
            self,
            "SageMakerBuildRole",
            assumed_by=iam.CompositePrincipal(
                iam.ServicePrincipal("sagemaker.
```

```
amazonaws.com")
                ),
                managed_policies=[
                    iam.ManagedPolicy.from_aws_managed_
policy_name("AmazonSageMakerFullAccess")
                ]
            )
    ...
```

12. The following code snippet shows how to create an S3 bucket to store the various assets created during the CI/CD pipeline execution:

```
    ...
            s3_bucket = s3.Bucket(
                self,
                "PipelineBucket",
                bucket_name=f"{repo_name}-{cdk.Aws.REGION}-
{cdk.Aws.ACCOUNT_ID}",
                removal_policy=cdk.RemovalPolicy.DESTROY,
                versioned=True
            )
            s3_bucket.grant_read_write(sagemaker_role)
            s3_bucket.grant_read_write(workflow_role)
    ...
```

13. As was the case with the SageMaker role, we also need to provide access to the name of the S3 bucket to the ML practitioners for usage outside of the pipeline. The following code snippet shows how we store the name of the S3 bucket in the SSM parameter store:

```
    ...
            s3_bucket_param = ssm.StringParameter(
                self,
                "PipelineBucketParameter",
                description="Pipeline Bucket Name",
                parameter_name="PipelineBucketName",
                string_value=s3_bucket.bucket_name
            )
    ...
```

14. Now that the S3 bucket has been created, we can copy the raw training dataset using the `BucketDeployment()` method:

```
...
        s3_deployment.BucketDeployment(
            self,
            "DeployData",
            sources=[
                s3_deployment.Source.asset(os.path.
join(os.path.dirname(__file__), '../artifacts/data'))
            ],
            destination_bucket=s3_bucket,
            destination_key_prefix="abalone_data/raw",
            retain_on_delete=False
        )
...
```

15. You will recall from *Chapter 4, Continuous Integration and Continuous Delivery (CI/CD) for Machine Learning*, that we created multiple CodeBuild projects. Each project correlated to a specific task within the ML process. Since the ML practitioners will be automating the entirety of the ML process as a Step Functions state machine, we now only need to define a single CodeBuild project to build the state machine artifact as a pipeline asset. The following code snippet shows how we define this single project and instantiate it as the `workflow_build` variable:

```
...
        workflow_build = codebuild.Project(
            self,
            "WorkflowBuildProject",
            project_name="WorkflowBuildProject",
            description="CodeBuild Project for Building
and Executing the ML Workflow",
            role=workflow_role,
            source=codebuild.Source.code_commit(
                repository=code_repo
            ),
            environment=codebuild.BuildEnvironment(
                build_image=codebuild.LinuxBuildImage.
STANDARD_5_0
```

```
                ),
                environment_variables={
                    "PIPELINE_NAME": codebuild.
BuildEnvironmentVariable(
                        value=repo_name
                    ),
                    "MODEL_NAME": codebuild.
BuildEnvironmentVariable(
                        value=model_name
                    ),
                    "BUCKET_NAME": codebuild.
BuildEnvironmentVariable(
                        value=s3_bucket.bucket_name
                    )
                }
            )
    ...
```

16. We also need to create an additional CodeBuild project to build out the model deployment parameters by executing the `deploy.py` script that we previously created. The following code snippet shows how to create the additional CodeBuild project and instantiate it by declaring the `deployment_build` parameter:

```
    ...
        deployment_build = codebuild.PipelineProject(
            self,
            "DeploymentBuild",
            project_name="DeploymentBuild",
            description="CodeBuild Project to Synthesize
    a SageMaker Endpoint CloudFormation Template",
            environment=codebuild.BuildEnvironment(
                build_image=codebuild.LinuxBuildImage.
STANDARD_5_0
            ),
            environment_variables={
                "BUCKET_NAME": codebuild.
BuildEnvironmentVariable(
                    value=s3_bucket.bucket_name
                ),
```

```
                         "PIPELINE_NAME": codebuild.
BuildEnvironmentVariable(
                            value=repo_name
                        ),
                         "MODEL_NAME": codebuild.
BuildEnvironmentVariable(
                            value=model_name
                        )
                    },
    ...
```

17. The build specification, or build instructions, for `deployment_build` has three phases, namely `install`, `build`, and `post_build`. The following code snippet shows the `install` phase, where we provide commands to install the AWS CDK and the relevant Python libraries required to create the endpoint deployment construct:

```
    ...
                            "install": {
                                "runtime-versions": {
                                    "python": 3.8,
                                    "nodejs": 12
                                },
                                "commands": [
                                    "echo Updating build
environment",
                                    "npm install aws-cdk@{}".
format(cdk_version),
                                    "python -m pip install
--upgrade pip",
                                    "python -m pip install -r
requirements.txt"
                                ]
                            },
    ...
```

18. After the relevant libraries have been installed, in the `install` phase, we synthesize the endpoint deployment construct and create an output of the resultant CloudFormation template, in JSON format, called `EndpointStack.template.json`. The following code snippet shows the build commands used in the `build` phase:

```
...
                    "build": {
                        "commands": [
                            "echo Synthesizing cdk
    template",
                            "npx cdk synth -o output"
                        ]
                    },
...
```

19. Once the CloudFormation template has been synthesized, the final phase of the build specification is to execute `deploy.py`. You will recall from the previous steps, the `deploy.py` file creates the `params.json` file to store the current CI/CD pipeline execution parameters. The following code snippet shows an example of the `post_build` phase:

```
...
                    "post_build": {
                        "commands": [
                            "python ./artifacts/
    scripts/deploy.py"
                        ]
                    }
                },
...
```

20. Now that all the required components and artifacts for the CI/CD pipeline have been defined, we can finally define the CI/CD pipeline itself. However, before we define the pipeline, we need to define output artifact variables. These variables correspond to the initial source code, as well as the various output files created by deployment_build. The following code snippet shows the main_source_output, model_source_output, and deployment_build_output variables being declared:

```
...
        main_source_output = codepipeline.Artifact()
        model_source_output = codepipeline.Artifact()
        deployment_build_output = codepipeline.
Artifact("DeploymentBuildOutput")
...
```

21. Now that we've declared the various source artifact, we can move onto defining the CI/CD pipeline using the Pipeline() method from the codepipeline CDK nodule. The pipeline has four stages, namely the Source stage, the Build stage, the Approval stage, and the Deploy stage. The following code snippet defines the Source stage. As you can see, within this stage, we declare the two branches of our CodeCommit repository, once for the pipeline CDK code (the main branch) and one for the model workflow artifacts (the model branch):

```
...
                    stage_name="Source",
                    actions=[
                        pipeline_actions.
CodeCommitSourceAction(
                            action_name="MainSource",
                            branch="main",
                            repository=code_repo,
                            output=main_source_output
                        ),
                        pipeline_actions.
CodeCommitSourceAction(
                            action_name="ModelSource",
                            branch="model",
                            repository=code_repo,
                            output=model_source_output
                        )
```

```
                ]
            ),
    ...
```

22. Within the Build stage, we define an action that calls the `workflow_build` CodeBuild project. You will recall that this is the CodeBuild project to create and execute the Step Functions state machine. The following code snippet declares the `BuildModel` stage action that references the `workflow_build` project:

```
    ...
                    codepipeline.StageProps(
                        stage_name="Build",
                        actions=[
                            pipeline_actions.CodeBuildAction(
                                action_name="BuildModel",
                                project=workflow_build,
                                input=model_source_output,
                                run_order=1
                            )
                        ]
                    ),
    ...
```

23. The penultimate stage of the CI/CD pipeline is the Approval stage. As you can see in the following code snippet, here we create a stage action called `EvaluationApproval` whereby we use the `ManualApprovalAction()` method from the `pipeline_actions` CDK module to add a manual approval step to the pipeline. It's at this point in the pipeline's execution that the process owners will verify that the model is ready to be deployed into production:

```
                    codepipeline.StageProps(
                        stage_name="Approval",
                        actions=[
                            pipeline_actions.
    ManualApprovalAction(
                                action_
    name="EvaluationApproval",
                                additional_information="Is
    the Model Ready for Production?"
                            )
```

```
                    ]
                ),
    ...
```

24. The final stage of the pipeline is where the trained model is deployed as a SageMaker hosted endpoint. It's at this stage that the previously defined `EndpointStack()` construct is deployed using the `DeployEndpoint` stage action. However, as you can see in the following code snippet, before the CDK construct can be used, we define a stage action called `DeploymentBuild` whereby we run the `deployment_build` CodeBuild project to synthesize the CloudFormation template, as well as the CloudFormation parameters file, needed to execute `CloudFormationCreateUpdateStackAction()`:

```
...
                    codepipeline.StageProps(
                        stage_name="Deploy",
                        actions=[
                            pipeline_actions.CodeBuildAction(
                                action_
name="DeploymentBuild",
                                project=deployment_build,
                                input=main_source_output,
                                outputs=[deployment_build_
output],
                                run_order=1
                            ),
                            pipeline_actions.
CloudFormationCreateUpdateStackAction(
                                action_name="DeployEndpoint",
                                stack_name="EndpointStack",
                                template_path=deployment_
build_output.at_path(
                                    "EndpointStack.template.
json"
                                ),
                                admin_permissions=True,
                                parameter_overrides={
                                    "ExecutionId":
deployment_build_output.get_param("params.json",
```

```
    "ExecutionId"),
                                "BucketName": deployment_
build_output.get_param("params.json", "BucketName"),
                            },
                            extra_inputs=[deployment_
build_output],
                            run_order=2
                        )
                    ]
                )
            ]
        )
    ...
```

Now that all the CDK constructs have been defined, we can put them all together and define the CDK application. The following steps will show you how to do this:

1. In the abalone-cicd-pipeline folder, open the app.py file for editing.

2. Delete the existing template code and add the following code to define the CDK application:

```
#!/usr/bin/env python3
import os
from aws_cdk import core as cdk
from abalone_cicd_pipeline.abalone_endpoint_stack import
EndpointStack
from abalone_cicd_pipeline.abalone_cicd_pipeline_stack
import PipelineStack
MODEL = "abalone"
CODECOMMIT_REPOSITORY = "abalone-cicd-pipeline"
CDK_VERSION = "2.3.0"
app = cdk.App()
EndpointStack(
    app,
    "EndpointStack",
    env=cdk.Environment(account=os.getenv("CDK_DEFAULT_
ACCOUNT"), region=os.getenv("CDK_DEFAULT_REGION")),
    model_name=MODEL,
    repo_name=CODECOMMIT_REPOSITORY
```

```
)
PipelineStack(
    app,
    CODECOMMIT_REPOSITORY,
    env=cdk.Environment(account=os.getenv("CDK_DEFAULT_
ACCOUNT"), region=os.getenv("CDK_DEFAULT_REGION")),
    model_name=MODEL,
    repo_name=CODECOMMIT_REPOSITORY,
    cdk_version=CDK_VERSION
)
app.synth()
```

3. Save and close the app.py file.

Since the CI/CD pipeline has been codified, we can go ahead and deploy it.

Deploying the CI/CD pipeline

Execute the following commands to deploy the CDK application and create the CI/CD pipeline:

1. Using the Terminal windows of the Cloud9 workspace, run the following commands to download the abalone training data from the UCI repository:

```
$ cd ~/environment/abalone-cicd-pipeline/ && mkdir -p
artifacts/data
$ wget -c -P artifacts/data https://archive.ics.uci.edu/
ml/machine-learning-databases/abalone/abalone.data
```

2. To ensure that all the code we've just created is committed to the CodeCommit repository, run the following commands to update these changes:

```
$ git add -A
$ git commit -m "Initial commit of Pipeline Artifacts"
$ git push --set-upstream origin main
```

3. Now, run the following command to deploy the CDK application:

```
$ cdk deploy abalone-cicd-pipeline
```

As we saw in the previous chapter, you can view the pipeline in the CodePipeline console (`https://console.aws.amazon.com/codesuite/codepipeline/`) by clicking on **Pipeline** in the console's navigation pane. *Figure 6.3* shows an example of what you might see:

Figure 6.3 – CodePipeline console

As you can see in *Figure 6.3*, **abalone-cicd-pipeline** has failed. If you click on the pipeline to open the details, you will see that the pipeline failed because there is no **ModelSource**. This is because the ML practitioner team hasn't created any model source artifacts yet. In the next chapter, we will work through creating the state machine artifacts using the Data Science SDK.

Summary

In this chapter, we re-factored the *Age Calculator* example from *Chapter 4, Continuous Integration and Continuous Delivery (CI/CD) for Machine Learning*, to further streamline the overall ML process by integrating the development teams and ML practitioner teams based on their areas of expertise.

For example, with this re-factored process, the development teams can now focus their expertise on building and developing the CI/CD components, while the ML practitioner teams can focus on codifying the ML process by using the Data Science SDK.

In the next chapter, we will switch personas to the ML practitioner team and review how they can codify the ML workflow as a Step Functions state machine.

7

Building the ML Workflow Using AWS Step Functions

In this chapter, we will continue from where we left off in *Chapter 6, Automating the Machine Learning Process Using AWS Step Functions*. You will recall from that chapter that the primary goal we are working toward achieving is to streamline the process gap that was originally highlighted in *Chapter 4, Continuous Integration and Continuous Delivery (CI/CD) for Machine Learning*—namely, to automate the handover of trained **machine learning (ML)** models from the ML practitioner teams to the development teams. Since we've already created **continuous integration/continuous delivery (CI/CD)** pipeline artifacts, as the application development engineers, the next step to achieving our goal is to provide the ML practitioner's contribution to further automate the **end-to-end (E2E)** process.

So, in this chapter, we are going to create a processing process that creates training and testing datasets, trains an ML model, and then evaluates the model's predictive quality, assessing whether it can be deployed into production. As you will see, the automated process will be codified as a CI/CD pipeline artifact using the **Amazon Web Services (AWS)** Step Functions Data Science **Software Development Kit (SDK)** for Python and developed from the perspective of the ML practitioner, without the need to upskill the development team members with capabilities outside their domain of expertise.

Once we've codified the E2E ML process as an AWS Step Functions state machine, we will continue to automate the *Age Calculator* use case by integrating the ML practitioner's modeling and workflow assets into the previously built CI/CD pipeline. To this end, we will be covering the following topics in this chapter:

- Building the state machine workflow
- Performing the integration test
- Monitoring the pipeline's progress

Technical requirements

To follow along with the code examples in this chapter, you will need the following:

- Web browser (for the best experience, it is recommended that you use Chrome or Firefox browsers).

- Access to the AWS account that we used in *Chapter 6, Automating the Machine Learning Process Using AWS Step Functions*.

- Access to the Cloud9 development environment we used in *Chapter 6, Automating the Machine Learning Process Using AWS Step Functions*.

- We will once again be working within the usage limits of the AWS Free Tier to avoid incurring unnecessary costs.

- Access to the SageMaker Studio environment we created in *Chapter 2, Automating Machine Learning Model Development Using SageMaker Autopilot*.

- Source code examples and Jupyter Notebooks are provided in the companion GitHub repository for this chapter (`https://github.com/PacktPublishing/Automated-Machine-Learning-on-AWS/tree/main/Chapter07`). The code examples should already be available in the Cloud9 development environment; if not, refer to the *Developing the application artifacts* section of *Chapter 4, Continuous Integration and Continuous Delivery (CI/CD) for Machine Learning*.

Building the state machine workflow

From the *Deploying the CI/CD pipeline* section of *Chapter 6, Automating the Machine Learning Process Using AWS Step Functions*, you will recall that we deployed a CI/CD pipeline to orchestrate the E2E ML process as a **Cloud Development Kit (CDK)** application. However, as you saw in *Figure 6.3*, the `abalone-cicd-pipeline` execution failed as there were no `ModelSource` artifacts.

Consequently, it's at this stage of the overall process that the ML practitioner must create these source artifacts to build the ML workflow, using the AWS Step Functions Data Science SDK for Python. We will therefore switch our perspective to that of the ML practitioner and build these source artifacts, using the SageMaker Studio **user interface** (**UI**).

Setting up the service permissions

Before we can begin defining the state machine workflow within a Jupyter Notebook, we need to assign additional access permissions for the SageMaker execution role to accommodate the Data Science SDK. According to the SDK documentation (`https://aws-step-functions-data-science-sdk.readthedocs.io/en/stable/readmelink.html#aws-permissions`), using the SDK in SageMaker Studio doesn't require any additional **Identity and Access Management** (**IAM**) permissions outside of those required by the Step Functions service. For example, if we were to use AWS Lambda functions within the workflow, we would need to add AWS Lambda permissions to the SageMaker execution role.

However, you may recall from *Chapter 2, Automating Machine Learning Model Development Using SageMaker Autopilot* that we added an extra inline IAM policy called `AdminAccess-InlinePolicy` to the SageMaker execution role. So, since the SageMaker execution role already has the necessary permissions to create and test the workflow, we can go ahead and actually build out the workflow using the Data Science SDK.

Creating an ML workflow

Since the Data Science SDK was primarily created to be executed within a Jupyter Notebook, we will use the following steps to codify the ML process as a notebook:

1. Using the **Amazon SageMaker** Management Console (`https://console.aws.amazon.com/sagemaker/home`), click the **SageMaker Domain** option in the left-hand navigation panel to open the **SageMaker Domain** dashboard.

2. Within the **SageMaker Domain** dashboard, click the **Launch app** dropdown and select the **Studio** option to launch the Studio UI.

> **Note**
>
> After opening the Studio UI, if you already see an `abalone-cicd-pipeline` folder in the left-hand Studio navigation panel, this means that the repository we cloned in *Chapter 5, Continuous Deployment of a Production ML Model* has not been deleted. Since we are using the same repository name for this chapter, simply right-click on the `abalone-cicd-pipeline` folder and then click **Delete** so that we can add the new repository we created in *Chapter 6, Automating the Machine Learning Process Using AWS Step Functions*.

3. Once the Studio UI has been launched, click on the **Git** icon in the left sidebar, and then click the **Clone a Repository** button, as illustrated in the following screenshot:

Figure 7.1 – Clone a Repository

4. In the **Clone a Repository** dialog window, enter the **HyperText Transfer Protocol Secure (HTTPS) Uniform Resource Locator (URL)** for the pipeline repository created by the development teams. The URL should be `https://git-codecommit.<AWS Region>.amazonaws.com/v1/repos/abalone-cicd-pipeline`, where `<AWS Region>` is the current region you are using. For example, if you are using `us-west-2` as the current region, then the URL would be `https://git-codecommit.us-west-2.amazonaws.com/v1/repos/abalone-cicd-pipeline`.

5. Click on the **CLONE** button to clone the pipeline repository.

6. Once the repository has been cloned, open the `abalone-cicd-pipeline` folder in the Studio navigation panel by double-clicking on the folder.

7. Now, click on the **Git** icon again to open the folder as the current repository.

8. Click on the drop-down arrow next to **Current Branch** and click the **New Branch** button, as illustrated in the following screenshot:

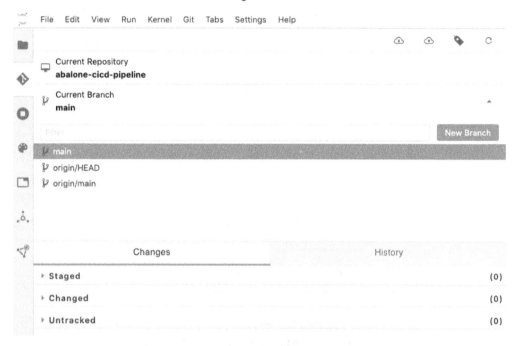

Figure 7.2 – New Branch

9. In the **Create a Branch** dialog, enter model as the new branch name and click **Create Branch**, as illustrated in the following screenshot:

Figure 7.3 – Create a Branch dialog

> **Note**
>
> For more information on cloning a Git repository in the Studio UI, see the SageMaker documentation (https://docs.aws.amazon.com/sagemaker/latest/dg/studio-tasks-git.html).

10. Now, go back to the abalone-cicd-pipeline folder and delete the existing contents.

11. Right-click inside the folder and select the **New Folder** menu option. Name the folder notebook and double-click to open it.

12. Create a new Jupyter Notebook by clicking the **File** menu, then clicking the **New** menu option, and then clicking on **Notebook**.

13. When prompted, make sure to select the **Python 3 (Data Science)** kernel and click the **Select** button.

14. Once the new notebook has been launched and the kernel has been started, we can go ahead and create our workflow, using the subsequent code.

> **Note**
>
> An example of the notebook is provided in the companion GitHub repository (`https://github.com/PacktPublishing/Automated-Machine-Learning-on-AWS/blob/main/Chapter07/Notebook/Abalone%20Step%20Functions%20Workflow%20Example.ipynb`). Since the following code examples show full Jupyter Notebook cells, as well as partial code snippets, it is recommended that you use this example notebook as a reference.

15. In the first code cell, enter the following code to install the necessary Data Science SDK and SageMaker SDK:

```
%%capture
!pip install stepfunctions==2.2.0 sagemaker==2.49.1
```

> **Note**
>
> We are hardcoding the versions of the `stepfunctions` and `sagemaker` libraries as these have been tested to work within the context of the example.

16. You will recall from the *Building the pipeline application artifacts* section of *Chapter 6, Automating the Machine Learning Process Using AWS Step Functions,* that we constructed the ML workflow orchestration as a CodeBuild project called `workflow_build`. Now, we will create a runtime process that will be executed within the CodeBuild environment by using the following code to instantiate the CodeBuild environment variables. These environment variables will make sense when we start using them later in the notebook; however, as you can see from the following code snippet, we are calling the **Systems Manager Parameter Store (SSM)** to fetch the **Simple Storage Service (S3)** bucket variable that was defined in the CDK application:

```
import os
import boto3
os.environ["MODEL_NAME"] = "abalone"
os.environ["PIPELINE_NAME"] = "abalone-cicd-pipeline"
os.environ["BUCKET_NAME"] = f"""{boto3.client("ssm").
get_parameter(Name="PipelineBucketName")["Parameter"]
["Value"]}"""
```

```
os.environ["DATA_PREFIX"] = "abalone_data"
os.environ["EPOCHS"] = "200"
os.environ["BATCH_SIZE"] = "8"
os.environ["THRESHOLD"] = "2.1"
```

17. In the next cell, we are going to build a custom cell magic. In previous examples, we've used a %%writefile magic to capture the cell contents to a file. However, the %%writefile magic does not execute the cell contents. Since we are building and testing a workflow creation script, the following code will create a custom magic called %%custom_writefile, whereby we capture the cell contents to a file, as well as run the contents:

```
from IPython.core.magic import register_cell_magic
@register_cell_magic
def custom_writefile(line, cell):
    print("Writing {}".format(line.split()[0]))
    with open(line.split()[0], "a") as f:
        f.write(cell)
    print("Running Cell")
    get_ipython().run_cell(cell)
```

> **Note**
>
> The %%custom_writefile magic is based on the examples provided in the official IPython documentation (https://ipython.readthedocs.io/en/stable/config/custommagics.html#defining-custom-magics). The only downside to using this methodology is that if we make a coding mistake in a particular cell, we have to delete any files that %%custom_writefile creates and start afresh from the beginning of the notebook.

18. Next, we create a folder called workflow, wherein we define a build script to create the workflow. Enter and execute the following code into a new cell:

```
!mkdir ../workflow
```

19. Now, we can start building our primary script, called main.py. In a new code cell, add the following code to start importing and capturing the various Python libraries to build the workflow:

```
%%custom_writefile ../workflow/main.py
import io
```

```
import os
import random
import time
import uuid
import boto3
import botocore
import zipfile
import json
from time import gmtime, strftime, sleep
from botocore.exceptions import ClientError
```

20. Next, we load the libraries required by the Data Science SDK, as follows:

```
%%custom_writefile ../workflow/main.py
import stepfunctions
from stepfunctions import steps
from stepfunctions.inputs import ExecutionInput
from stepfunctions.steps import (
    Chain,
    ChoiceRule,
    ModelStep,
    ProcessingStep,
    TrainingStep,
    TuningStep,
    TransformStep,
    Task,
    EndpointConfigStep,
    EndpointStep,
    LambdaStep
)
from stepfunctions.template import TrainingPipeline
from stepfunctions.template.utils import replace_
parameters_with_jsonpath
from stepfunctions.workflow import Workflow
```

21. And lastly, we load the libraries from the SageMaker SDK, as follows:

```
%%custom_writefile ../workflow/main.py
import sagemaker
from sagemaker.tensorflow import TensorFlow
from sagemaker.tuner import IntegerParameter,
ContinuousParameter, HyperparameterTuner
from sagemaker import get_execution_role
from sagemaker.amazon.amazon_estimator import get_image_
uri
from sagemaker.processing import ProcessingInput,
ProcessingOutput, Processor
from sagemaker.s3 import S3Uploader
from sagemaker.sklearn.processing import SKLearnProcessor
```

22. After importing the relevant Python libraries, we use the following code to define connections to the various AWS services used through the workflow:

```
%%custom_writefile ../workflow/main.py
sagemaker_session = sagemaker.Session()
region = sagemaker_session.boto_region_name
role = get_execution_role()
sfn_client = boto3.client("stepfunctions")
lambda_client = boto3.client("lambda")
codepipeline_client = boto3.client("codepipeline")
ssm_client = boto3.client("ssm")
```

23. Now that we have loaded the required libraries, we are going to define some helper functions. The first helper function we will need is the `get_execution_role()` function. The following code defines this function to get the SSM parameter for the workflow execution role that was created as part of the CDK application:

```
%%custom_writefile ../workflow/main.py
def get_workflow_role():
    try:
        response = ssm_client.get_parameter(
            Name="WorkflowRoleParameter",
        )
        return response["Parameter"]["Value"]
    except ClientError as e:
```

```
        error_message = e.response["Error"]["Message"]
        print(error_message)
        raise Exception(error_message)
```

24. The following code defines a second function, called `update_lambda()`. This function will update any AWS Lambda function's code if the function already exists:

```
%%custom_writefile ../workflow/main.py
def update_lambda(name, zip_name):
    lambda_client.update_function_code(
        FunctionName=name,
        ZipFile=open(zip_name, mode="rb").read(),
        Publish=True
    )
```

25. The next helper function is called `get_lambda()`. This function takes any defined AWS Lambda code, zips it, and creates a new Lambda function. If the Lambda function already exists, `get_lambda()` will call `update_lambda()` to update the existing Lambda function with the updated code. The code is illustrated in the following snippet:

```
%%custom_writefile ../workflow/main.py
def get_lambda(name, bucket, description):
    print("Creating Lambda Package ")
    zip_name = f"../artifacts/{name}.zip"
    lambda_src = f"../artifacts/{name}.py"
    z = zipfile.ZipFile(zip_name, mode="w")
    z.write(lambda_src, arcname=lambda_src.split("/")
[-1])
    z.close()
    print("Uploading Lambda Package to S3 ")
    S3Uploader.upload(
        local_path=zip_name,
        desired_s3_uri=f"s3://{bucket}/lambda",
    )
    try:
        print(f"Creating Lambda Function '{name}' …")
        lambda_client.create_function(
            FunctionName=name,
```

```
            Runtime="python3.8",
            Role=get_workflow_role(),
            Handler=f"{name}.lambda_handler",
            Code={
                "S3Bucket": bucket,
                "S3Key": f"lambda/{name}.zip"
            },
            Description=description,
            Timeout=120,
            MemorySize=128
        )
    except ClientError as e:
        print(f"Lambda Function '{name}' already exists,
re-creating ...")
        update_lambda(name, zip_name)
    return name
```

26. The final helper function we will define is called get_execution_id(). This function calls CodePipeline to get the **identifier** (**ID**) of the current execution. You will recall that for this, we will be versioning the workflow execution, and thus pipeline assets, based on the current execution ID. If there is no execution ID, then we will use the current time as a versioning ID. The code is illustrated in the following snippet:

```
%%custom_writefile ../workflow/main.py
def get_execution_id(name=None):
    try:
        response = codepipeline_client.get_pipeline_
state(name=name)
        for stage in response["stageStates"]:
            if stage["stageName"] == "Build":
                for action in stage["actionStates"]:
                    if action["actionName"] ==
"BuildModel":
                        return stage["latestExecution"]
["pipelineExecutionId"]
    except KeyError:
        return strftime('%Y%m%d%H%M%S', gmtime())
```

> **Note**
>
> *SageMaker expects unique names for each job, trained model, and endpoint. If these names are not unique then execution will fail. So, if we are unit testing the workflow code and not running it as part of the CI/CD pipeline, then we need to supply a unique version to the SageMaker job.*

27. Now that we have created helper functions, we can proceed to declaring additional parameters that are specific to the workflow itself, as well as unique to a workflow execution. The following code defines unique workflow parameters:

```
%%custom_writefile ../workflow/main.py
execution_id = get_execution_id(name=os.
environ["PIPELINE_NAME"])
model = os.environ["MODEL_NAME"]
data_prefix = os.environ["DATA_PREFIX"]
model_prefix = execution_id
bucket_name = os.environ["BUCKET_NAME"]
model_name = f"{model}-{execution_id}"
training_job_name = f"{model}-TrainingJob-{execution_id}"
preprocessing_job_name = f"{model}-ProcessingJob-
{execution_id}"
evaluation_job_name = f"{model}-EvaluationJob-{execution_
id}"
deeplearning_container_image = f"763104351884.dkr.ecr.
{region}.amazonaws.com/tensorflow-training:2.5.0-cpu-
py37-ubuntu18.04-v1.0"
```

28. Next, we define execution parameters as an `ExecutionInput()` schema. The schema defines the type of parameters that will be provided to start a workflow execution. The code is illustrated in the following snippet:

```
%%custom_writefile ../workflow/main.py
execution_input = ExecutionInput(
    schema={
        "ModelName": str,
        "PreprocessingJobName": str,
        "TrainingJobName": str,
        "EvaluationProcessingJobName": str
    }
)
```

29. The final set of parameters we define specifies the data configuration. Here, we define an S3 location to get the raw abalone data, as well as an S3 location for the data once it has been processed:

```
%%custom_writefile ../workflow/main.py
s3_bucket_base_uri = f"s3://{bucket_name}"
input_data = os.path.join(s3_bucket_base_uri,  data_
prefix, "raw/abalone.data")
output_data = os.path.join(s3_bucket_base_uri, data_
prefix)
preprocessed_training_data = os.path.join(output_data,
"input", "training")
preprocessed_testing_data = f"{output_data}/testing"
model_data_s3_uri = f"{s3_bucket_base_uri}/{model_
prefix}/{training_job_name}/output/model.tar.gz"
output_model_evaluation_s3_uri = f"{s3_bucket_base_uri}/
{model_prefix}/{training_job_name}/evaluation"
```

Now that we've defined the required global variables, helper functions, and overall workflow parameters, the next stage is to codify the workflow itself. The following steps will walk you through how to create the steps that make up the workflow:

1. The first step in the workflow is to process the raw abalone data as a SageMaker processing job. However, before defining the processing step, we need to provide SageMaker with a processing script. The following code cell creates an `artifacts` folder to store the various script artifacts:

```
!mkdir ../artifacts
```

2. Now, we use the following code to capture the `preprocessing.py` processing artifact script:

```
%%writefile ../artifacts/preprocessing.py
import os
import pandas as pd
import numpy as np
prefix = "/opt/ml"
processing_path = os.path.join(prefix, "processing")
preprocessing_input_path = os.path.join(processing_path,
"input")
preprocessing_output_path = os.path.join(processing_path,
```

```
    "output")
if __name__ == "__main__":
    print("Preprocessing Data")
    column_names = ["sex", "length", "diameter",
"height", "whole_weight", "shucked_weight", "viscera_
weight", "shell_weight", "rings"]
    data = pd.read_csv(os.path.join(preprocessing_input_
path, "abalone.data"), names=column_names)
    y = data.rings.values.reshape(len(data), 1)
    del data["rings"]
    print("Creating Catagorical Features")
    data = pd.get_dummies(data).to_numpy()
    X = np.concatenate((y, data), axis=1)
    print("Splitting Data into Training, Validation and,
Test Datasets")
    training, validation, testing = np.split(X,
[int(.8*len(X)), int(.95*len(X))])
    pd.DataFrame(training).to_csv(os.path.
join(preprocessing_output_path, "training/training.csv"),
header=False, index=False)
    pd.DataFrame(validation).to_csv(os.path.
join(preprocessing_output_path, "training/validation.
csv"), header=False, index=False)
    pd.DataFrame(testing).to_csv(os.path.
join(preprocessing_output_path, "testing/testing.csv"),
header=False, index=False)
    print("Done!")
```

Note

As you can see, this script contains the same data processing methodology
we've been using throughout the book for the abalone dataset.

---

3. Now that we have the script for the processing job, we can define a workflow step definition to call SageMaker and execute the processing job as a task within the workflow. In the following code snippet, we start defining `processing_step` as a `ProcessingStep()` state machine:

```
. . .
%%custom_writefile ../workflow/main.py
processing_step = ProcessingStep(
    "Pre-process Data",
. . .
```

4. Next, we specify the type of processing job as `SKLearnProcessor()` and the type of compute resources to use for the processing job, as follows:

```
. . .
        processor=SKLearnProcessor(
            framework_version="0.23-1",
            role=role,
            instance_type="ml.m5.xlarge",
            instance_count=1,
            max_runtime_in_seconds=1200,
        ),
        job_name=execution_input["PreprocessingJobName"],
. . .
```

5. As the following code snippet shows, we now specify the location of the input data for the processing job, as well as the location of the `processing.py` script we created in *Step 2*:

```
. . .
        inputs=[
            ProcessingInput(
                source=input_data,
                destination="/opt/ml/processing/input",
                input_name="input"
            ),
            ProcessingInput(
                source=sagemaker_session.upload_data(
                    path="../artifacts/preprocessing.py",
```

```
                bucket=bucket_name,
                key_prefix=os.path.join(data_prefix,
 "code")
            ),
            destination="/opt/ml/processing/input/code",
            input_name="code"
        )
    ],
...
```

6. Once inputs have been defined, we can define outputs. In the following code snippet, we define output locations for both the training and testing data. These datasets will be stored in the S3 bucket we defined in *Step 7* earlier:

```
...
    outputs=[
        ProcessingOutput(
            source="/opt/ml/processing/output/training",
            destination=os.path.join(output_data, "input",
 "training"),
            output_name="training"
        ),
        ProcessingOutput(
            source="/opt/ml/processing/output/testing",
            destination=os.path.join(output_data,
 "testing"),
            output_name="testing"
        )
    ],
...
```

7. The final part of the processing_step state machine, as shown in the following code snippet, is to specify the preprocessing.py script as the execution entry point to processing_step:

```
...
    container_entrypoint=["python3", "/opt/ml/processing/
 input/code/preprocessing.py"],
 )
...
```

8. Once the data has been processed, we can move on to the next step of the workflow, where we train the model. In this step, we will call SageMaker to run a training job. Before we can define the workflow step, we need to provide SageMaker with the model training code as an artifact. The following code snippet shows how a `model_training.py` artifact is created. As you can see, we define a process to train a TensorFlow model using the same methodology as the previous examples:

```python
...
if __name__ == "__main__":
    print(f"Tensorflow Version: {tf.__version__}")
    column_names = ["rings", "length", "diameter",
"height", "whole weight", "shucked weight", "viscera
weight", "shell weight", "sex_F", "sex_I", "sex_M"]
    parser = argparse.ArgumentParser()
    parser.add_argument('--epochs', type=int, default=2)
    parser.add_argument('--batch-size', type=int,
default=8)
    parser.add_argument('--model-dir', type=str,
default=os.environ['SM_MODEL_DIR'])
    parser.add_argument('--training', type=str,
default=os.environ['SM_CHANNEL_TRAINING'])
    args, _ = parser.parse_known_args()
    epochs = args.epochs
    batch_size = args.batch_size
    training_path = args.training
    model_path = args.model_dir
    train_data = pd.read_csv(os.path.join(training_path,
'training.csv'), sep=',', names=column_names)
    val_data = pd.read_csv(os.path.join(training_path,
'validation.csv'), sep=',', names=column_names)
    train_y = train_data['rings'].to_numpy()
    train_X = train_data.drop(['rings'], axis=1).to_
numpy()
    val_y = val_data['rings'].to_numpy()
    val_X = val_data.drop(['rings'], axis=1).to_numpy()
    train_X = preprocessing.normalize(train_X)
    val_X = preprocessing.normalize(val_X)
    network_layers = [Dense(64, activation="relu",
kernel_initializer="normal", input_dim=10), Dense(64,
```

```
activation="relu"), Dense(1, activation="linear")]
    model = Sequential(network_layers)
    model.compile(optimizer='adam', loss='mse',
metrics=['mae', 'accuracy'])
    model.summary()
    model.fit(train_X, train_y, validation_data=(val_X,
val_y), batch_size=batch_size, epochs=epochs,
shuffle=True, verbose=1)
    model.save(os.path.join(model_path, 'model.h5'))
    model_version = 1
    export_path = os.path.join(model_path, str(model_
version))
    tf.keras.models.save_model(model, export_path,
overwrite=True, include_optimizer=True, save_format=None,
signatures=None, options=None)
...
```

9. Now that the training artifact is created, we define a `training_step` workflow step to define an instance of the `TrainingStep()` workflow task. As part of the task, we specify the location of the training script, the hyperparamaters, the job name, the type of compute resources to use, and the S3 location of the processed training data. The code is illustrated in the following snippet:

```
%%custom_writefile ../workflow/main.py
training_step = TrainingStep(
    "Model Training",
    estimator=TensorFlow(
        entry_point='../artifacts/model_training.py',
        role=role,
        hyperparameters={
            'epochs': int(os.environ['EPOCHS']),
            'batch-size': int(os.environ['BATCH_SIZE']),
        },
        train_instance_count=1,
        train_instance_type='ml.m5.xlarge',
        framework_version='2.4',
        py_version="py37",
        script_mode=True,
        output_path=os.path.join(s3_bucket_base_uri,
```

```
model_prefix)
    ),
    data={"training": sagemaker.inputs.
TrainingInput(preprocessed_training_data, content_
type="csv")},
    job_name=execution_input["TrainingJobName"],
    wait_for_completion=True,
)
```

10. After the model has been trained, we need to evaluate whether or not it qualifies as a production-grade model. We will define a SageMaker processing job to execute the evaluation, and as we did with the data processing task, we define a script artifact called evaluate.py. This artifact will load the trained model, plus the testing dataset, and capture the model inference output to an evaluation.json file in S3. The following code creates an artifact and loads the necessary Python libraries for the evaluation:

```
%%writefile ../artifacts/evaluate.py
import json
import os
import tarfile
import pandas as pd
import numpy as np
import tensorflow as tf
from tensorflow import keras
from tensorflow.keras.models import Sequential
from tensorflow.keras.layers import Dense
from tensorflow.keras.optimizers import Adam
from sklearn import preprocessing
```

11. Next, we define a Python function to load and compile the trained TensorFlow model, as follows:

```
%%writefile -a ../artifacts/evaluate.py


def load_model(model_path):
    model = tf.keras.models.load_model(os.path.
join(model_path, 'model.h5'))
    model.compile(optimizer='adam', loss='mse')
    return model
```

12. Now, we append another function to the artifact to capture the inferences from the loaded model and store the results in S3, as follows:

```
%%writefile -a ../artifacts/evaluate.py
def evaluate_model(prefix, model):
    column_names = ["rings", "length", "diameter",
"height", "whole weight", "shucked weight",
                    "viscera weight", "shell weight",
"sex_F", "sex_I", "sex_M"]
    input_path = os.path.join(prefix, "processing/
testing")
    output_path = os.path.join(prefix, "processing/
evaluation")
    predictions = []
    truths = []
    test_df = pd.read_csv(os.path.join(input_path,
"testing.csv"), names=column_names)
    y = test_df['rings'].to_numpy()
    X = test_df.drop(['rings'], axis=1).to_numpy()
    X = preprocessing.normalize(X)
    for row in range(len(X)):
        payload = [X[row].tolist()]
        result = model.predict(payload)
        print(result[0][0])
        predictions.append(float(result[0][0]))
        truths.append(float(y[row]))
    report = {
        "GroundTruth": truths,
        "Predictions": predictions
    }
    with open(os.path.join(output_path, "evaluation.
json"), "w") as f:
        f.write(json.dumps(report))
```

13. Finally, we append the main program to execute the evaluation, as follows:

```
%%writefile -a ../artifacts/evaluate.py
if __name__ == "__main__":
    print("Extracting model archive ...")
```

```
prefix = "/opt/ml"
model_path = os.path.join(prefix, "model")
tarfile_path = os.path.join(prefix, "processing/
model/model.tar.gz")
with tarfile.open(tarfile_path) as tar:
    tar.extractall(path=model_path)
print("Loading Trained Model ...")
model = load_model(model_path)
print("Evaluating Trained Model ...")
evaluate_model(prefix, model)
print("Done!")
```

14. As with the data processing step, we use the following code to define another `ProcessingStep()` workflow to execute the `evaluation.py` artifact, as follows:

```
%%custom_writefile ../workflow/main.py
evaluation_step = ProcessingStep(
    "Model Evaluation",
    processor=Processor(
        image_uri=deeplearning_container_image,
        instance_count=1,
        instance_type="ml.m5.xlarge",
        role=role,
        max_runtime_in_seconds=1200
    ),
    job_name=execution_
input["EvaluationProcessingJobName"],
    inputs=[
        ProcessingInput(
            source=preprocessed_testing_data,
            destination="/opt/ml/processing/testing",
            input_name="input"
        ),
        ProcessingInput(
            source=model_data_s3_uri,
            destination="/opt/ml/processing/model",
            input_name="model"
        ),
```

```
ProcessingInput(
    source=sagemaker_session.upload_data(
        path="../artifacts/evaluate.py",
        bucket=bucket_name,
        key_prefix=os.path.join(data_prefix,
"code")
    ),
    destination="/opt/ml/processing/input/code",
    input_name="code"
)
],
outputs=[
    ProcessingOutput(
        source="/opt/ml/processing/evaluation",
        destination=output_model_evaluation_s3_uri,
        output_name="evaluation"
    )
],
container_entrypoint=["python3", "/opt/ml/processing/
input/code/evaluate.py"]
)
```

Note

SageMaker processing jobs natively support processing data using **Apache Spark** or the `scikit-learn` Python libraries. Since we are evaluating a TensorFlow model, which isn't natively supported, we leverage the TensorFlow training **deep learning (DL)** container image (`https://github.com/aws/deep-learning-containers/blob/master/available_images.md#general-framework-containers`), using the `image_uri` parameter to perform the model evaluation within the `ProcessingStep()` state machine.

---

15. Since we've captured the inference result in the `evaluation.json` file, we need to assess the results against an evaluation metric. To do this, we will use an AWS Lambda function. The following code snippet shows the `lambda_handler()` definition, as an artifact called `analyze_results.py`:

```
...
def lambda_handler(event, context):
```

```python
    logger.debug("## Environment Variables ##")
    logger.debug(os.environ)
    logger.debug("## Event ##")
    logger.debug(event)
    s3 = boto3.client("s3")
    if ("Bucket" in event):
        bucket = event["Bucket"]
    else:
        raise KeyError("S3 'Bucket' not found in Lambda
event!")
    if ("Key" in event):
        key = event["Key"]
    else:
        raise KeyError("S3 'Key' not found in Lambda
event!")
    logger.info("Downloading evlauation results file
...")
    json_file = json.loads(s3.get_object(Bucket = bucket,
Key = key)['Body'].read())
    logger.info("Analyzing Model Evaluation Results ...")
    y = json_file["GroundTruth"]
    y_hat = json_file["Predictions"]
    summation = 0
    for i in range (0, len(y)):
        squared_diff = (y[i] - y_hat[i])**2
        summation += squared_diff
    rmse = math.sqrt(summation/len(y))
    logger.info("Root Mean Square Error: {}".
format(rmse))
    logger.info("Done!")
    return {
        "statusCode": 200,
        "Result": rmse,
    }
...
```

16. To run the Lambda function as a step within the workflow, we define a
    `LambdaStep()` function and use the helper functions to create a Lambda,
    as follows:

```
%%custom_writefile ../workflow/main.py
analyze_results_step = LambdaStep(
    "Analyze Evaluation Results",
    parameters={
        "FunctionName": get_lambda(
            "analyze_results",
            bucket_name,
            "Analyze the results from the Model
Evaluation"
        ),
        "Payload": {
            "Bucket": bucket_name,
            "Key": f"""{model_prefix}/{training_job_
name}/evaluation/evaluation.json"""
        }
    }
)
```

17. The final task within the workflow is to register the trained model as a SageMaker
    model. This is the model that will be deployed as a hosted endpoint during the CD
    phase of the CI/CD pipeline. The following code creates a `ModelStep()` function
    and points to the trained model from the `TrainingStep()` workflow task:

```
%%custom_writefile ../workflow/main.py
register_model_step = ModelStep(
    "Register Trained Model",
    model=training_step.get_expected_model(),
    model_name=execution_input["ModelName"],
    instance_type="ml.m5.large"
)
```

18. Now that we have all the steps of the workflow, we need to put them together and define the flow of the state machine. To do this, we will work backward from the various end results of workflow execution to build up the workflow. In the following code snippet, we define the resultant state of a failed workflow and connect each step to the failed state, if they should fail:

```
%%custom_writefile ../workflow/main.py
workflow_failed_state = stepfunctions.steps.states.Fail(
    "ML Workflow Failed",
cause="SageMakerProcessingJobFailed"
)
```

```
catch_state = stepfunctions.steps.states.Catch(error_
equals=["States.TaskFailed"], next_step=workflow_failed_
state)
processing_step.add_catch(catch_state)
training_step.add_catch(catch_state)
evaluation_step.add_catch(catch_state)
analyze_results_step.add_catch(catch_state)
register_model_step.add_catch(catch_state)
```

```
If the trained model fails the evaluation—or, in other
words, is above the
threshold we establish for a production-grade model—the
workflow should
also fail. In the next code snippet, we define a failure
state for the model exceeding the evaluation criteria:
```

```
%%custom_writefile ../workflow/main.py
```

```
threshold_fail_state = stepfunctions.steps.states.Fail(
    "Model Evaluation Exceeds Threshold"
)
```

```
Along with declaring the final failure states of the
workflow, we also need to create a final state whereby
the model's evaluation determines it to be below the
evaluation threshold, and therefore a production-grade
model. The following code snippet defines this Pass()
state:
```

```
%%custom_writefile ../workflow/main.py
```

```
threshold_pass_state = stepfunctions.steps.states.Pass(
    "Model Evaluation Below Threshold"
)
```

19. To determine whether or not the model evaluation is above or below the evaluation criteria, we define a `Choice()` state and configure a `ChoiceRule()` function to determine whether the output of the `analyze_results_step` task is less than the `THRESHOLD` variable, as follows:

```
%%custom_writefile ../workflow/main.py
check_threshold_step = steps.states.Choice(
    "Threshold Evaluation Check"
)
threshold_rule = steps.choice_rule.ChoiceRule.
NumericLessThan(
    variable=analyze_results_step.output()['Payload']
['Result'],
    value=float(os.environ["THRESHOLD"])
)
check_threshold_step.add_choice(rule=threshold_rule,
next_step=threshold_pass_state)
check_threshold_step.default_choice(next_step=threshold_
fail_state)
```

20. We've just created all of our steps and states of the ML workflow, as well as the supporting artifacts the various steps will use. The final part of creating our workflow is to put them all together. The following code chains the various steps together and creates a workflow graph:

```
%%custom_writefile ../workflow/main.py
_graph = Chain(
    [
        processing_step,
        training_step,
        register_model_step,
        evaluation_step,
        analyze_results_step,
        check_threshold_step
    ]
)
```

21. We now have our workflow defined, using the Data Science SDK. If we refer to *Figure 6.2* of *Chapter 6*, *Automating the Machine Learning Process Using AWS Step Functions*, we can see that the next part of the process for the ML practitioner to perform is to unit test, and validate the code works. The following code creates a workflow called `abalaone-workflow-unit-test`, and then executes it:

```
ml_workflow = Workflow(ml_workflow
    name="abalone-workflow-unit-test",
    definition=ml_workflow_graph,
    role=get_workflow_role(),
)
ml_workflow.create()
execution = ml_workflow.execute(
    inputs={
        "ModelName": model_name,
        "PreprocessingJobName": preprocessing_job_name,
        "TrainingJobName": training_job_name,
        "EvaluationProcessingJobName": evaluation_job_
name,
    }
)
execution_output = execution.get_output(wait=True)
```

> **Note**
>
> Executing a unit test on the workflow will incur additional AWS resource costs outside of the Free Tier. You can forego the previous step to avoid additional charges.

22. Based on the success of the unit-test procedure, the final addition to the main.py script is to capture the process to execute a production state machine. The following code will create an abalone-workflow production workflow and provide the execution-specific parameters for a production execution:

```
%%writefile -a ../workflow/main.py
print("Creating ML Workflow")
ml_workflow = Workflow(
    name="abalone-workflow",
    definition=ml_workflow_graph,
    role=get_workflow_role(),
)
try:
    print("Creating Step Functions State Machine")
    ml_workflow.create()
except sfn_client.exceptions.StateMachineAlreadyExists:
    print("Found Existing State Machine, Updating the
State Machine definition")
else:
    ml_workflow.update(ml_workflow_graph)
    time.sleep(120)
print("Executing ML Workflow State Machine")
ml_workflow.execute(
    inputs={
        "ModelName": model_name,
        "PreprocessingJobName": preprocessing_job_name,
        "TrainingJobName": training_job_name,
        "EvaluationProcessingJobName": evaluation_job_
name
    }
)
```

23. This completes the ML practitioner's contribution to the refactored solution. All that's left to do is to commit these changes to the repository. To do this, click on the **Git** icon.

24. Click the plus (+) icon for both the **Changed** and **Untracked** sections, to move the changes into the **Staged** section.

25. In the **Summary (required)** field, provide a summary of these changes, by entering Initial commit of Workflow Artifacts, as illustrated in the following screenshot:

Figure 7.4 – Staged changes

26. Click the **Commit** button to commit the changes to the model branch.

27. Once the changes have been committed, click **Git** from the menu bar, and select **Push to Remote**.

> **Note**
> If prompted, provide your email address and name.

By checking in this code, we, as the ML practitioner, have completed our contribution to the refactored solution. We have used the Data Science SDK to codify an ML workflow and create a state machine, without having to use the Amazon States Language. The following screenshot shows a graphical representation of our ML workflow:

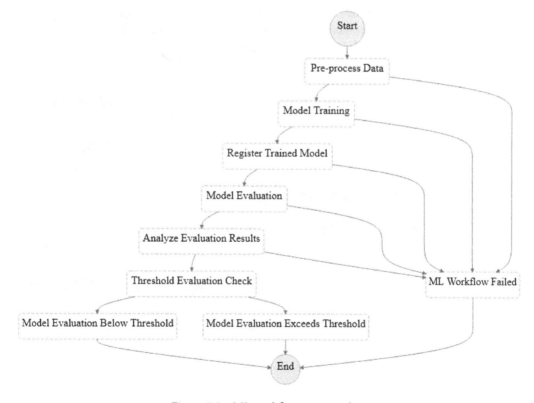

Figure 7.5 – ML workflow state machine

However, even though the state machine code has been checked in, the overall CI/CD pipeline still won't execute as we have not yet defined the integration between the pipeline and the state machine. In the next section, we will once again perform the final integration, from the perspective of the development engineer.

Performing the integration test

To finalize the CI/CD pipeline for release, we need to integrate the code that the ML practitioner submitted into the build process. We do this by providing the build instructions to the CodeBuild stage by creating a `buildspec.yml` file.

> **Note**
>
> You can find a complete copy of the `buildspec.yml` file, for your reference, in the companion GitHub repository (`https://github.com/PacktPublishing/Automated-Machine-Learning-on-AWS/blob/main/Chapter07/Files/buildspec.yml`).

The following steps will walk you through the integration process, performed from the perspective of the development engineer:

1. Using the Cloud9 environment, run the following command within the Terminal window to pull the latest changes that the ML practitioner made:

    ```
    $ cd ~/environment/abalone-cicd-pipeline/ && git pull
    ```

2. Change to the `model` branch by running the following command:

    ```
    $ git checkout model
    ```

3. Right-click on the `abalone-cicd-pipeline` folder in the navigation panel and select **New File**.

4. Name the file buildspec.yml and double-click on it for editing.

5. Add the following code to declare instructions for loading the necessary Python libraries and executing the main.py script:

```
version: 0.2
env:
  variables:
    DATA_PREFIX: abalone_data
    EPOCHS: 200
    BATCH_SIZE: 8
    THRESHOLD: 2.1

phases:
  install:
    runtime-versions:
      python: 3.8
    commands:
      - printenv
      - echo "Updating Build Environment"
      - apt-get update
      - python -m pip install --upgrade pip
      - python -m pip install --upgrade boto3 awscli
sagemaker==2.49.1 stepfunctions==2.2.0
  build:
    commands:
      - echo Build started on 'date'
      - echo "Creating ML Workflow "
      - |
        sh -c """
        cd workflow/
        python main.py
        """
  post_build:
    commands:
      - echo "Build Completed"
```

6. Save the file.

7. Within the Terminal window, run the following command to add the `buildspec.yml` file to the file repository file index:

    ```
    $ git add -A
    ```

8. Commit the integration to the CodeCommit repository by running the following command:

    ```
    $ git commit -m "Add Integration Artifacts"
    ```

9. Now, push the changes to the repository with the following command:

    ```
    $ git push
    ```

We have now integrated the workflow into the CI/CD pipeline, and by committing these changes, we have also created a pipeline release. In the next section, we will monitor the pipeline.

Monitoring the pipeline's progress

Monitoring the pipeline execution is done through the CodePipeline console. In the web browser, open the AWS CodePipeline Management Console (`https://console.aws.amazon.com/codesuite/codepipeline/home`), and then click on the name of the pipeline—`abalone-cicd-pipeline`. The following screenshot depicts the pipeline execution:

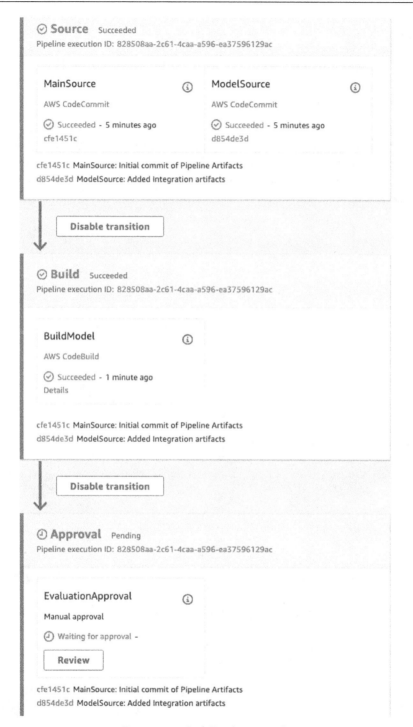

Figure 7.6 – CodePipeline console

If you compare *Figure 7.6* with the pipeline in *Figure 5.4* of *Chapter 5, Continuous Deployment of a Production ML Model*, the first thing you will notice is that the **Build** stage has been significantly compressed to a action called **BuildModel**. This is because we are offloading the ML modeling process to the Step Functions state machine, instead of capturing the modeling process into the pipeline itself.

To review the progress of the state machine in a new web browser tab, open the AWS Step Functions Management Console (`https://console.aws.amazon.com/states`) and select the `abalone-workflow` state machine. You will see a list of executions. Click on the latest execution to review its progress. The following screenshot shows the **succeeded** execution:

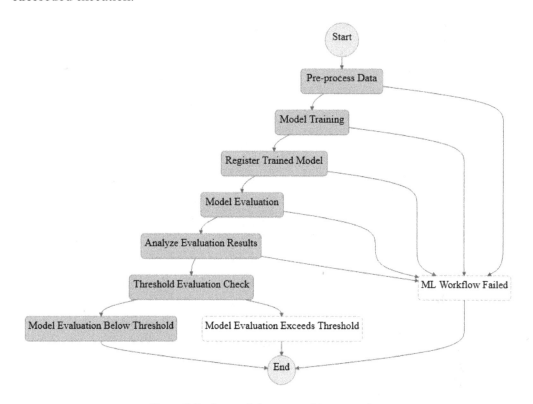

Figure 7.7 – Succeeded state machine execution

As you can see from *Figure 7.7*, the workflow has successfully completed, producing a trained ML model that is below the pre-established evaluation criteria. To verify the evaluation criteria, click on the **Analyze Evaluation Results** step of the **Graph inspector**, and then click on the **Step output** tab. The following screenshot shows an example result of the model evaluation:

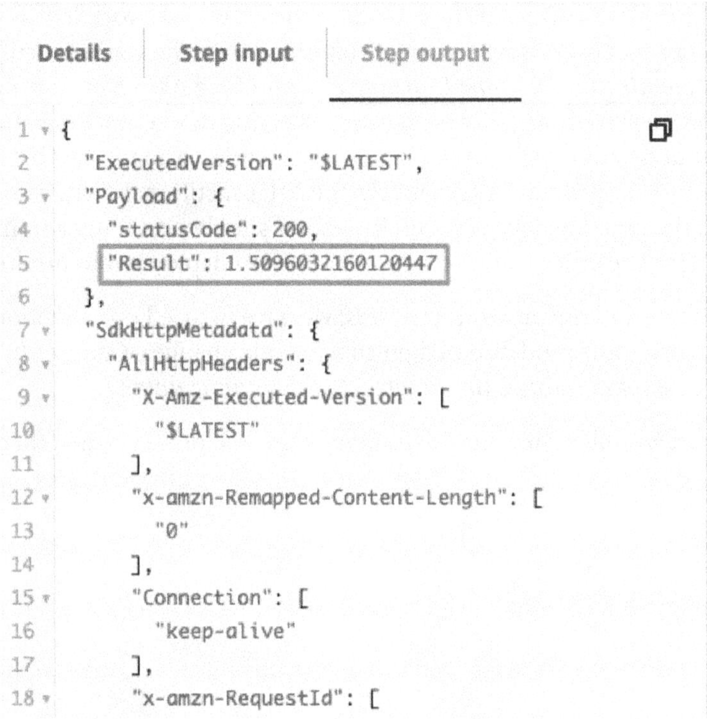

```
Details        Step input       Step output


1 ▼ {
2      "ExecutedVersion": "$LATEST",
3 ▼   "Payload": {
4        "statusCode": 200,
5        "Result": 1.5096032160120447
6      },
7 ▼   "SdkHttpMetadata": {
8 ▼     "AllHttpHeaders": {
9 ▼       "X-Amz-Executed-Version": [
10          "$LATEST"
11        ],
12 ▼      "x-amzn-Remapped-Content-Length": [
13          "0"
14        ],
15 ▼      "Connection": [
16          "keep-alive"
17        ],
18 ▼      "x-amzn-RequestId": [
```

Figure 7.8 – Analyze Evaluation Results: Step output

This completes the CI phase of the pipeline, and we can once again approve the model for deployment.

> **Note**
>
> See the *Executing the automated ML model deployment* and *Cleanup* sections of *Chapter 5, Continuous Deployment of a Production ML Model*, to see how to approve a model and continue with the CD phase of a pipeline.

Once the **DeployEndpoint** action of CodePipeline is complete, we have a production model that can be integrated into the *Age Calculator* application and serve abalone age predictions.

Summary

In this chapter, we continued refactoring the *Age Calculator* example that we started in *Chapter 6, Automating the Machine Learning Process Using AWS Step Functions*, to further streamline the overall ML automation process, using AWS Step Functions.

Not only have we seen how ML practitioner teams can tighten their integration with the development (or platform) teams by providing the entire ML workflow as a CI/CD pipeline artifact, but we also saw how—when combined with the codified artifacts created in *Chapter 6, Automating the Machine Learning Process Using AWS Step Functions*—each team can focus on their specific area of expertise. Now, the development teams don't have to upskill their understanding of how the ML process works to adapt the CI/CD pipeline to accommodate the ML process. Alternatively, the ML practitioner team can contribute their expertise to the pipeline development, instead of simply providing a trained ML model and expecting the other teams to figure out how to deploy it into production.

However, in both this and the previous chapters, we have focused our attention on releasing production-grade models based on source code changes. *Conversely, how does the CI/CD process adapt to changes (or updates) to the training data?*

In the next chapter, we will review how to automate the ML process when there are source data changes.

Section 4: Optimizing a Data-Centric Approach to Automated Machine Learning

This section introduces you to what a data-centric ML process is, how it differs from a code-centric approach, and the services typically used for this methodology, namely, Apache Airflow and Amazon Managed Workflows for Apache Airflow.

This section comprises the following chapters:

8
Automating the Machine Learning Process Using Apache Airflow

When building an ML model, there is a fundamental principle that all ML practitioners are aware of; namely, an ML model is only as robust as the data on which it was trained. In the previous four chapters, we have primarily focused on automating the ML process using a source code-centric mechanism. In other words, we applied a DevOps methodology of Continuous Integration and Continuous Deployment to automate the ML process by supplying the model source code, tuning parameters, and the ML workflow source code. Any changes to these artifacts would trigger a release change process of the CI/CD pipeline.

However, we also supplied static abalone data, downloaded from the UCI Machine Learning Repository, as a source artifact, but we never made any changes to this data. So, using a typical DevOps methodology, the data artifact is static and therefore won't trigger a change release of the CI/CD process.

Accordingly, data becomes the key differentiator between applying a DevOps methodology to automate the ML process, versus an MLOps strategy. For a successful MLOps strategy, we basically need to provide the ability to automate the ML process when the data changes. In essence, just as the ML automation process is triggered when source code is added or updated, we also need to trigger the automation process when existing data is updated, or new data is added.

This then begs the question; **how can we automate the ML process when we have new data?**

To answer this question, in this chapter, we will focus on automating the ML process on AWS, using a data-centric approach. The overall objective of this chapter is to duplicate the foundation that we've established in both *Chapter 4, Continuous Integration and Continuous Delivery (CI/CD) for Machine Learning*, as well as *Chapter 6, Automating the Machine Learning Process Using AWS Step Functions*, to automate the *Age Calculator* example, with new training data. We will be accomplishing this by means of the following topics:

- Introducing Apache Airflow
- Introducing Amazon MWAA
- Using Airflow to process the abalone dataset
- Configuring the MWAA prerequisites
- Configuring the MWAA environment

Technical requirements

We will use the following resources in this chapter:

- A web browser (for the best experience, it is recommended that you use Chrome or Firefox).
- Access to the AWS account that you used in *Chapter 7, Building the ML Workflow Using AWS Step Functions*.
- Access to the Cloud9 development environment we used in *Chapter 7, Building the ML Workflow Using AWS Step Functions*.
- We will once again be working within the usage limits of the AWS Free Tier to avoid incurring unnecessary costs.

- Source code examples are provided in the companion GitHub repository for this chapter (`https://github.com/PacktPublishing/Automated-Machine-Learning-on-AWS/tree/main/Chapter07`). The code examples should already be available in the Cloud9 development environment; if not, refer to the *Developing the application artifacts* section in *Chapter 4, Continuous Integration and Continuous Delivery (CI/CD) for Machine Learning*.

Introducing Apache Airflow

Data for ML model training can come from various sources, such as databases, data warehouses, or even data lakes. These data repositories store data in a wide variety of different data formats. For example, data may be stored as unstructured objects, as in the case of image, video, or sound files. Objects may be stored as semi-structured data, such as JSON data that doesn't conform to a standardized tabular schema. In the case of relational databases, or data warehouses, the data is stored in an organized and structured format, but it may have multiple different types of schemas.

To make matters worse, some datasets can be very large, often terabytes, or even petabytes in size, where joining, merging, and transforming the data, often referred to as **Extract, Transform and Load (ETL)** processes, requires large compute clusters, such as **Hadoop** and **Apache Spark** clusters. AWS provides infrastructure resources and dedicated services to scale these big data workloads in the form of **AWS Glue** (a managed ETL service) and **Amazon Elastic Map Reduce**, or **EMR** (big data platform).

However, performing ETL tasks on these different types of big data, and their varying sources, often requires daisy-chaining multiple separate ETL tasks together as part of an orchestrated workflow, where the data output from one ETL task becomes the input to another ETL task, and so on. As you can imagine, creating such a workflow can be a daunting task.

So, to simplify the process, many data engineers rely on **Apache Airflow** (`https://airflow.apache.org/`), a platform that allows them to programmatically construct, execute, and manage these potentially complex data workflows. The Airflow platform comprises three key components, namely:

- A web-based management interface
- A scheduler, responsible for scheduling and coordinating the resources to execute the various steps, or tasks, within the workflow
- Multiple workers to execute the code for each specific task within the workflow

To use the Airflow platform, a data engineer creates a codified representation of a workflow in the form of a **Directed Acyclic Graph (DAG)**. *Figure 8.1* shows an example of what an Airflow DAG looks like:

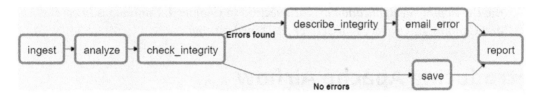

Figure 8.1 – Airflow DAG example

> **Note**
> *Figure 8.1* is made available under the Apache 2.0 license and can be referenced in the Airflow GitHub repository (`https://github.com/apache/airflow/blob/main/docs/apache-airflow/img/edge_label_example.png`).

As you can see from *Figure 8.1*, the DAG is made up of various sequential, or directed, tasks that are programmatically defined as a Python construct. Once the DAG is submitted to the scheduler, the scheduler coordinates its execution by assigning each task to a worker. Each worker, in turn, processes the code for the individual task to which it was assigned.

So, using Airflow significantly simplifies the data engineer's task or orchestrates these complex data transformation tasks. However, having yet another platform to manage now adds additional complexity for the infrastructure and operations teams, as now these teams must manage the big data processing platforms (such as Hadoop and Spark), as well as the Airflow platform.

How can the platform management tasks also be simplified?

To answer this question, we will explore **Amazon Managed Workflows for Apache Airflows (MWAA)** next.

Introducing Amazon MWAA

Managing big data platforms isn't typically part of the ML practitioner's portfolio of tasks. Oftentimes, the ML practitioner and data engineering teams rely on the infrastructure and operations teams to manage these platforms.

Including Airflow as part of the big data infrastructure means that these platform teams must now manage additional compute resources, orchestrate their deployment, update software and operating system patches, and monitor these resources to ensure that they are constantly addressing workload scaling requirements and other **Service-Level Agreements (SLAs)**.

AWS offers multiple big data managed services, such as EMR and Glue, to help offload these management tasks from the platform teams and, in November 2020, AWS launched Amazon MWAA to help offload the management of the Airflow platform. With MWAA, the platform teams can run a highly available and scalable Airflow cluster without having to individually provision, update, and monitor the Web UI server, scheduler, or even the worker nodes. This means that the ML practitioner and data engineering teams can focus on developing the data workflow without relying on the platform teams.

To illustrate just how MWAA can work in practice, we are going to leverage the service for the *Age Calculator* use case.

Using Airflow to process the abalone dataset

To set the scene, you will recall from *Chapter 1*, *Getting Started with Automated Machine Learning on AWS*, that the *ACME Fishing Logistics* company uses an outdated dataset, found in the UCI Machine Learning Repository, to train the ML model. The ML practitioners have found that since an ML model is only as good as the data it's trained on, they can tweak and tune the model as much as they want, but without newer data, the production model can't be improved upon.

To resolve this problem, ACME has hired an external company to survey abalone catches and supply daily updates of the surveyed dataset. This means that the already tuned ML model can be retrained on fresh data, and thus be further optimized. This also means that the data engineering teams need to orchestrate a process, or data pipeline, to merge the original dataset with the new survey data and supply the new training, validation, and testing dataset to a new model training pipeline, all using the MWAA service.

Let's see how the various ACME teams would approach this task.

Configuring the MWAA prerequisites

Before we can launch the MWAA service, there are a few prerequisites that need to be addressed, namely:

- MWAA requires access to an S3 bucket where the DAGs are stored.

- MWAA needs to access a `requirements.txt` file, also stored on S3, to load any unique Python libraries that the workers would need to execute their assigned tasks.

- Although not required by MWAA, we need to also configure various IAM roles to access backend services such as Glue and SageMaker.

- We also need to provide the artifacts that the various backend services would require. For example, we need to provide ETL scripts in order for the Glue service to execute.

In the following steps, we will provide these prerequisites as a CDK application:

1. Log in to the same AWS account you've been using in the previous chapter and open the AWS Cloud9 console (`https://console.aws.amazon.com/cloud9`).

2. In the **Your environments** section, click the **Open IDE** button for the **MLOps-IDE** development environment.

> **Note**
>
> If you've been following along up to this point, you should already have the MLOps-IDE environment configured, along with version 2.3.0 of the AWS CDK. If not, please refer to the *Preparing the development environment* section in *Chapter 4, Continuous Integration and Continuous Delivery (CI/CD) for Machine Learning*.

3. Next, we will create a CodeCommit repository to store the various codified artifacts for the entire solution:

```
$ cd ~/environment
$ aws codecommit create-repository --repository-name
abalone-data-pipeline --repository-description "Automated
ML on AWS using Managed Workflows for Apache Airflow"
```

4. Now we can capture the repository's URI for cloning:

```
$ CLONE_URL=$(aws codecommit get-repository --repository-
name abalone-data-pipeline --query "repositoryMetadata.
cloneUrlHttp" --output text)
```

5. Run the following command to clone the repository:

```
$ git clone $CLONE_URL
```

6. Run the following commands to initialize the CDK application within the new repository:

```
$ cd abalone-data-pipeline && cdk init --language python
$ git add -A
$ git commit -m "Started CDK Project"
$ git branch main
$ git checkout main
$ source .venv/bin/activate
```

7. Next, we will install the necessary development libraries by running the following commands:

```
$ python -m pip install -U pip pylint boto3
sagemaker==2.49.1 apache-airflow
$ pip install -r requirements.txt
```

8. Now that we have the relevant libraries installed, we can start defining the various data pipeline resources. Using the Cloud9 navigation panel, on the left-hand side, expand the abalone_data_pipeline folder and double-click on the abalone_data_pipeline_stack.py file for editing.

> **Note**
>
> Examples of the CDK code can be found in the companion GitHub repository for this chapter (https://github.com/PacktPublishing/ Automated-Machine-Learning-on-AWS/tree/main/ Chapter08/cdk).

9. Delete the template code and add the following to import the necessary
    CDK libraries:

```
import os
import aws_cdk.aws_codecommit as codecommit
import aws_cdk.aws_codebuild as codebuild
import aws_cdk as cdk
import aws_cdk.aws_s3 as s3
import aws_cdk.aws_ssm as ssm
import aws_cdk.aws_s3_deployment as s3_deployment
import aws_cdk.aws_iam as iam
import aws_cdk.aws_glue as glue
import aws_cdk.aws_lambda as lambda_
import aws_cdk.aws_events_targets as targets
from constructs import Construct
```

10. Next, we initialize the `DataPipelineStack` class by adding the following code:

```
class DataPipelineStack(cdk.Stack):
    def __init__(self, scope: Construct, id: str, *,
airflow_environment_name: str=None, model_name: str=None,
repo_name: str=None, **kwargs) -> None:
        super().__init__(scope, id, **kwargs)
```

11. The first construct we will build is a reference to the CodeCommit
    repository. Adding the following code registers the CodeCommit
    repository as a CDK construct:

```
        code_repo = codecommit.Repository.from_
repository_name(
            self,
            "SourceRepository",
            repository_name=repo_name
        )
```

12. Now, we create the S3 bucket to store the relevant data and store the bucket name as
    an SSM parameter so that it can be referenced in the Airflow DAG:

> **Note**
>
> Since this is a data-centric solution, we will store all relevant assets for the workflow in a dedicated *data* bucket, as opposed to the *pipeline* bucket that we've used in previous examples.

```
Data_bucket = s3.Bucket(
    self,
    "AirflowDataBucket",
    bucket_name=f"{model_name}-data-{cdk.Aws.
REGION}-{cdk.Aws.ACCOUNT_ID}",
    block_public_access=s3.BlockPublicAccess.
BLOCK_ALL,
    auto_delete_objects=True,
    removal_policy=cdk.RemovalPolicy.DESTROY,
    versioned=True
)
ssm.StringParameter(
    self,
    "DataBucketParameter",
    description="Airflow Data Bucket Name",
    parameter_name="AirflowDataBucket",
    string_value=data_bucket.bucket_name
)
```

13. Next, we create a SageMaker role, which allows the Airflow workflow to initiate SageMaker API calls. We need to ensure that this role has access to the data bucket and is referenceable in the Airflow DAG. So, we also store the role ARN as an SSM parameter:

```
sagemaker_role = iam.Role(
    self,
    "SageMakerBuildRole",
    assumed_by=iam.CompositePrincipal(
        iam.ServicePrincipal("sagemaker.
amazonaws.com")
    ),
    managed_policies=[
        iam.ManagedPolicy.from_aws_managed_
policy_name("AmazonSageMakerFullAccess")
```

```
        ]
    )
    data_bucket.grant_read_write(sagemaker_role)
    ssm.StringParameter(
        self,
        "SageMakerRoleParameter",
        description="SageMaker Role ARN",
        parameter_name="SageMakerRoleARN",
        string_value=sagemaker_role.role_arn
    )
```

14. In the previous chapter, we created an AWS Lambda function to analyze the results from the ML model evaluation. This was done during the ML workflow build process. Since we are building out the various resources for the Airflow workflow, we are going to codify the Lambda function here. The function is also granted access to read the evaluation results file in the data bucket and is stored as an SSM parameter:

```
    analyze_results_lambda = lambda_.Function(
        self,
        "AnalyzeResults",
        handler="index.lambda_handler",
        runtime=lambda_.Runtime.PYTHON_3_8,
        code=lambda_.Code.from_asset(os.path.join(os.
path.dirname(__file__), "../artifacts/lambda/analyze_
results")),
        memory_size=128,
        timeout=cdk.Duration.seconds(60)
    )
    data_bucket.grant_read(analyze_results_lambda)
    ssm.StringParameter(
        self,
        "AnalyzeResultsParameter",
        description="Analyze Results Lambda Function
Name",
        parameter_name="AnalyzeResultsLambda",
```

```
                string_value=analyze_results_lambda.function_
name
            )
```

15. Next, we will use the following code to create the necessary resources to process the training, validation, and testing datasets. As already mentioned, we need to scale the data processing to handle potentially large datasets. So, to streamline this process, we will leverage the AWS Glue ETL service. The first resource that Glue requires is an IAM role with the necessary permissions to the data bucket:

```
        glue_role = iam.Role(
            self,
            "GlueRole",
            assumed_by=iam.CompositePrincipal(
                iam.ServicePrincipal("glue.amazonaws.
com")
            ),
            managed_policies=[
                iam.ManagedPolicy.from_aws_managed_
policy_name("service-role/AWSGlueServiceRole")
            ]
        )
        data_bucket.grant_read_write(glue_role)
```

16. We can now create a Glue Catalog to store references to the new abalone data:

```
        glue_catalog = glue.CfnDatabase(
            self,
            "GlueDatabase",
            catalog_id=cdk.Aws.ACCOUNT_ID,
            database_input=glue.CfnDatabase.
DatabaseInputProperty(
                name=f"{model_name}_new"
            )
        )
```

17. To populate the Glue Catalog with the new abalone data, use the following code to create a Glue Crawler. The crawler will crawl the new data in our data bucket and append it to the Glue Catalog:

```
glue_crawler = glue.CfnCrawler(
    self,
    "GlueCrawler",
    name=f"{model_name}-crawler",
    role=glue_role.role_arn,
    database_name=glue_catalog.ref,
    targets={
        "s3Targets": [
            {
                "path": f"s3://{data_bucket.
bucket_name}/{model_name}_data/new/"
            }
        ]
    }
)
```

18. We also need to store the crawler's name as an SSM parameter so that it can be referenced in the Airflow workflow:

```
ssm.StringParameter(
    self,
    "GlueCrawlerParameter",
    description="Glue Crawler Name",
    parameter_name="GlueCrawler",
    string_value=glue_crawler.name
)
```

19. Once the new data has been added to the Glue Catalog, we can create the Glue Job that reads this data, merges it with the original abalone dataset, and performs the necessary data preprocessing tasks to make the entire dataset ready for model training:

```
glue_job = glue.CfnJob(
    self,
    "GlueETLJob",
    name=f"{model_name}-etl-job",
    description="AWS Glue ETL Job to merge new +
raw data, and process training data",
    role=glue_role.role_arn,
    glue_version="2.0",
    execution_property=glue.CfnJob.
ExecutionPropertyProperty(
        max_concurrent_runs=1
    ),
    command=glue.CfnJob.JobCommandProperty(
        name="glueetl",
        python_version="3",
        script_location=f"s3://{data_bucket.
bucket_name}/airflow/scripts/preprocess.py"
    ),
    default_arguments={
        "--job-language": "python",
        "--GLUE_CATALOG": glue_catalog.ref,
        "--S3_BUCKET": data_bucket.bucket_name,
        "--S3_INPUT_KEY_PREFIX": f"{model_name}_
data/raw/abalone.data",
        "--S3_OUTPUT_KEY_PREFIX": f"{model_name}_
data",
        "--TempDir": f"s3://{data_bucket.bucket_
name}/glue-temp"
    },
    allocated_capacity=5,
    timeout=10
)
```

20. Since the Airflow workflow will be calling the Glue Job, we also store the job name as an SSM parameter:

```
ssm.StringParameter(
    self,
    "GlueJobParameter",
    description="Glue Job Name",
    parameter_name="GlueJob",
    string_value=glue_job.name
)
```

21. To ensure that the original abalone dataset is also available to the Glue ETL Job, we use the following code to create an S3 bucket deployment that uploads the *raw* dataset to S3:

```
s3_deployment.BucketDeployment(
    self,
    "DeployData",
    sources=[
        s3_deployment.Source.asset(os.path.
join(os.path.dirname(__file__), "../artifacts/data"))
    ],
    destination_bucket=data_bucket,
    destination_key_prefix=f"{model_name}_data/
raw",
    retain_on_delete=False
)
```

22. Finally, we create a CodeBuild project that allows the Airflow DAG to be continuously updated. Although we are building a data-centric workflow, we also need to ensure that any updates, or changes, to the codified workflow itself can automatically be applied. The following code snippet instantiates the code_deployment variable as a codebuild.Project():

```
...
code_deployment = codebuild.Project(
    self,
    "CodeDeploymentProject",
    project_name="CodeDeploymentProject",
    description="CodeBuild Project to Copy
```

```
    Airflow Artifacts to S3",
                source=codebuild.Source.code_commit(
                    repository=code_repo
                ),
                environment=codebuild.BuildEnvironment(
                    build_image=codebuild.LinuxBuildImage.
    STANDARD_5_0
                ),
                environment_variables={
                    "DATA_BUCKET": codebuild.
    BuildEnvironmentVariable(
                        value=data_bucket.bucket_name
                    )
                },
    ...
```

23. The CodeBuild project has three phases that make up the BuildSpec, or build instructions, namely, the install, build, and post_build phases. The following code snippet shows the install phase, where the latest version of the AWS CLI is installed:

```
    ...
                    "install": {
                        "runtime-versions": {
                            "python": 3.8
                        },
                        "commands": [
                            "printenv",
                            "echo 'Updating Build
    Environment'",
                            "python -m pip install
    --upgrade pip",
                            "python -m pip install
    --upgrade boto3 awscli"
                        ]
                    },
    ...
```

24. The following code snippet shows the AWS CLI commands that update the Airflow assets. During the `build` phase of the CodeBuild project, we `sync` the Airflow workflow assets to S3 so that once these have been copied to the data bucket, the MWAA scheduler will automatically import the new DAG code:

```
...
          "build": {
              "commands": [
                  "echo 'Deploying Airflow
Artifacts to S3'",
                  "cd artifacts",
                  "aws s3 sync airflow
s3://${DATA_BUCKET}/airflow"
              ]
          },
...
```

25. To ensure that the CodeBuild project has the appropriate access to synchronize the Airflow assets to S3, the following code snippet grants the `code_deploy` role read and write access to `data_bucket`:

```
...
        data_bucket.grant_read_write(code_deployment.
    role)
...
```

26. To ensure that the workflow changes are applied, we create an event trigger that starts the CodeBuild project every time any changes are committed to the source repository:

```
        code_repo.on_commit(
            "StartDeploymentProject",
            target=targets.CodeBuildProject(code_
    deployment)
        )
```

27. Save and close the `abalone_data_pipeline_stack.py` file.

28. Now that we have defined the necessary resources as a CDK construct, we can define the CDK application to deploy these resources. Using the workspace navigation panel, open the `app.py` file in the `abalone-data-pipeline` folder and delete the existing template code.

> **Note**
> A full example of the `app.py` code can be found in the companion GitHub repository for this chapter (`https://github.com/PacktPublishing/Automated-Machine-Learning-on-AWS/tree/main/Chapter08/cdk`).

29. Delete the existing template code and add the following code to define the CDK application:

```python
#!/usr/bin/env python3
import os
import aws_cdk as cdk
from abalone_data_pipeline.abalone_data_pipeline_stack import DataPipelineStack
MODEL = "abalone"
CODECOMMIT_REPOSITORY = "abalone-data-pipeline"
app = cdk.App()
DataPipelineStack(
    app,
    CODECOMMIT_REPOSITORY,
    env=cdk.Environment(account=os.getenv("CDK_DEFAULT_ACCOUNT"), region=os.getenv("CDK_DEFAULT_REGION")),
    model_name=MODEL,
    repo_name=CODECOMMIT_REPOSITORY,
    airflow_environment_name=f"{MODEL}-airflow-environment"
)
app.synth()
```

30. Save and close the `app.py` file.

31. Before we can deploy the CDK application for our workflow resources, we need to download the original abalone dataset from the UHCI repository. Run the following commands in the Cloud9 terminal:

```
$ cd ~/environment/abalone-data-pipeline/
$ mkdir -p artifacts/data
$ wget -c -P artifacts/data https://archive.ics.uci.edu/ml/machine-learning-databases/abalone/abalone.data
```

32. The next artifact is the source code for the `analyze_results` Lambda function. Run the following command to create the necessary folders:

```
$ cd ~/environment/abalone-data-pipeline/
$ mkdir -p artifacts/lambda/analyze_results
```

33. To define the Lambda function handler, we can reuse the code we created in *Chapter 7, Building the ML Workflow Using AWS Step Functions*. Run the following command to copy the `index.py` file from the already cloned companion GitHub repository:

```
$ cp ~/environment/src/Chapter08/lambda/analyze_results/
index.py ~/environment/abalone-data-pipeline/artifacts/
lambda/analyze_results/
```

> **Note**
>
> If you are unfamiliar with how the `analyze_results` Lambda function assesses the model evaluation results against the **Root Mean Squared Error (RMSE)** evaluation metric, you can review the code in the *Creating the ML workflow* section of *Chapter 7, Building the ML Workflow Using AWS Step Functions*.

34. The last artifact we need to configure is the `evaluate.py` file. You will recall from *Chapter 7, Building the ML Workflow Using AWS Step Functions*, that this script is executed as a SageMaker processing job to evaluate the trained model's performance on the testing dataset. Run the following commands to create the artifact folder and reuse the `evaluate.py` file provided in the companion GitHub repository:

```
$ cd ~/environment/abalone-data-pipeline/
$ mkdir -p artifacts/airflow/scripts
$ cp ~/environment/src/Chapter08/airflow/scripts/
evaluate.py ~/environment/abalone-data-pipeline/
artifacts/airflow/scripts/
```

> **Note**
>
> Should you need to re-familiarize yourself with the `evaluate.py` code, you can refer to the *Creating the ML workflow* section of *Chapter 7, Building the ML Workflow Using AWS Step Functions*.

35. Since we have all the necessary artifacts for the CDK application, we can go ahead and deploy it by running the following command in the terminal window:

```
cdk deploy
```

> **Note**
>
> The CDK should take approximately 5 minutes to deploy, and you can track the progress in the CloudFormation console (`https://console.aws.amazon.com/cloudformation/home`).

36. Once the resource stack has been deployed, we can add the initial Airflow artifacts. These artifacts are needed before we can deploy the MWAA environment. To do this, run the following commands to set up the artifact source folders:

```
$ cd ~/environment/abalone-data-pipeline/
$ mkdir -p artifacts/airflow/dags
```

37. The first Airflow artifact we will define is the `.airflowignore` file. This is a useful file for adding any DAG files that we want the Airflow scheduler to ignore. Run the following command to create this file:

```
$ touch artifacts/airflow/dags/.airflowignore
```

38. Next, we define the `requirements.txt` file. This file specifies the various Python dependencies that the Airflow DAG will need to install on the workers. Using the Cloud9 navigation panel, right-click on the `airflow` folder and select **New File**. Name the file `requirements.txt` and open it for editing.

> **Note**
>
> For more information on the best practices for managing Python dependencies in MWAA, see the MWAA documentation (`https://docs.aws.amazon.com/mwaa/latest/userguide/best-practices-dependencies.html`).

39. In the `requirements.txt` file, add the following dependencies:

```
sagemaker==2.49.1
s3fs==0.5.1
boto3>=1.17.4
```

> **Note**
>
> The only reason we are using version 2.49.1 of the SageMaker SDK is to ensure conformity with the ML experiment that the ML practitioner conducted in the previous chapter. It is good practice to keep the Python dependency versions constant with any source code provided by the data engineer or ML practitioner.

40. Save and close the `requirements.txt` file.

41. Since we have the necessary artifacts required for deploying the MWAA environment, we can run the following commands to commit these changes to the CodeCommit repository and have the CodeBuild project automatically update them in the data bucket:

```
$ git add -A
$ git commit -m "Initial commit of workflow artifacts"
$ git push --set-upstream origin main
```

The update should take about a minute to complete. We can view the output from the CodeBuild project in the console (`https://console.aws.amazon.com/codesuite/codebuild/projects`), selecting **CodeDeploymentProject** in the **Build projects** dashboard. As you can see, we now have a mechanism for deploying new and updated DAGs to MWAA.

Thus, we can move on to the next part of the process; deploying and configuring the MWAA environment.

Configuring the MWAA environment

Now that the necessary resources and prerequisites have been deployed, we can go ahead and provision the MWAA environment. The following steps will walk you through this procedure:

1. Open the MWAA console (`https://console.aws.amazon.com/mwaa/home`) in a new browser tab and click the **Create environment** button.

2. On the **Specify details** page, scroll down to the **DAG code in Amazon S3** section, and click the **Browse S3** button to our data bucket.

3. In the **Choose S3 bucket** window, check the radio button next to the bucket called **abalone-data-<REGION>-<ACCOUNT ID>** and then click **Choose**.

> **Note**
> Make sure that **<REGION>** and **<ACCOUNT ID>** in the bucket name match your environment.

4. Clicking **Choose** will return you to the **Specify details** page. On this page, click the **Browse S3** button under the **DAGs folder** section to open the **Choose DAG path in S3** window. *Figure 8.2* shows an example of this window:

Figure 8.2 – Choose DAG path in S3 window

5. As you can see from *Figure 8.2*, click the `airflow` folder to open it.

6. Once the `airflow` folder is open, select the radio button next to the `dags` folder, as shown in *Figure 8.3*:

Figure 8.3 – Selecting the dags folder

7. As you can see from *Figure 8.3*, once you have selected the dags folder, click the **Choose** button to return to the previous screen.

8. Now, click the **Browser S3** button under the **Requirements file – optional** section to choose the location of the requirements.txt file.

9. Repeat steps 5 and 6, but this time, click the radio button next to the requirements.txt file, as shown in *Figure 8.4*:

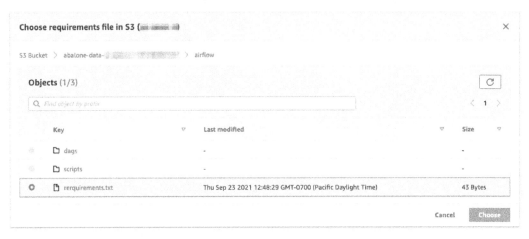

Figure 8.4 – Selecting the requirements.txt file

10. As you can see from *Figure 8.4*, once the requirements.txt file has been selected, click on the **Choose** button to return to the previous screen.

11. On the **Specify details** screen, click on the **Next** button, which will take you to the **Configure advanced settings** screen.

12. Under the **Networking** section of the **Configure advanced settings** screen, click on the **Create MWAA VPC** button to create a dedicated **Virtual Private Cloud** (**VPC**) for MWAA. This will launch the **Quick create stack** CloudFormation console in a new browser tab.

13. Leave all the fields as their defaults and click the **Create stack** button. The stack should take around 2 minutes to build, and once the status registers as **CREATE_ COMPLETED**, go back to the browser tab hosting the MWAA console.

14. Under the **Networking** section, click the refresh button and then, using the **Choose VPC** dropdown, select the VPC called **MWAA-VPC**, as shown in *Figure 8.5*:

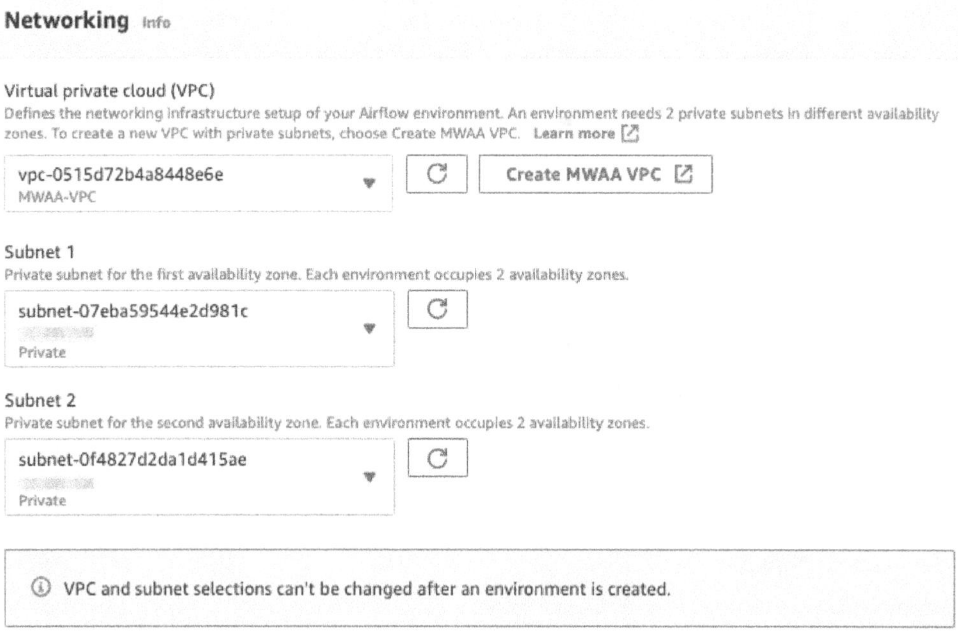

Networking Info

Virtual private cloud (VPC)

Defines the networking infrastructure setup of your Airflow environment. An environment needs 2 private subnets in different availability zones. To create a new VPC with private subnets, choose Create MWAA VPC. **Learn more** [↗]

| vpc-0515d72b4a8448e6e | ▼ | ⟳ | Create MWAA VPC [↗] |
| MWAA-VPC | | | |

Subnet 1

Private subnet for the first availability zone. Each environment occupies 2 availability zones.

| subnet-07eba59544e2d981c | ▼ | ⟳ |
| Private | | |

Subnet 2

Private subnet for the second availability zone. Each environment occupies 2 availability zones.

| subnet-0f4827d2da1d415ae | ▼ | ⟳ |
| Private | | |

ⓘ VPC and subnet selections can't be changed after an environment is created.

Figure 8.5 – MWAA VPC

15. As you can see from *Figure 8.5*, selecting the **MWAA-VPC** automatically populates the **Subnet 1** and **Subnet 2** fields with the correct private subnets.

16. Scroll down to the **Web server access** section and click on the radio button next to the **Public network (no additional setup)** option.

17. Leave the rest of the fields at their defaults and then click the **Next** button.

> **Note**
> Take note of the role name that is automatically created in the **Permissions** section. We will be assigning additional permissions for AWS services to this role.

18. Review the MWAA environment configuration and click the **Create environment** button.

> **Note**
> Deploying the MWAA environment will incur additional AWS usage costs that exceed the Free Tier usage. For more information on MWAA pricing, refer to the product pricing documentation (`https://aws.amazon.com/managed-workflows-for-apache-airflow/pricing/`).

19. While the environment is being provisioned, open the IAM console (`https://console.aws.amazon.com/iamv2/home?#/home`) in a new browser tab.

20. In the left-hand navigation panel of the IAM console, click on **Roles**.

21. In the **Roles** dashboard, scroll down until you see the IAM role that was created during the MWAA setup and then click on it.

22. In the Role **Summary** dashboard, click on the **Attach policies** button.

23. Using the search bar in the **Attach permissions** screen, search for and select the **AmazonS3FullAccess**, **AWSLambda_FullAccess**, **AmazonSSMFullAccess**, **AWSGlueConsoleFullAccess**, and **AmazonSageMakerFullAccess** policies.

24. Click the **Attach policies** button.

25. The **Summary** screen should resemble *Figure 8.6*:

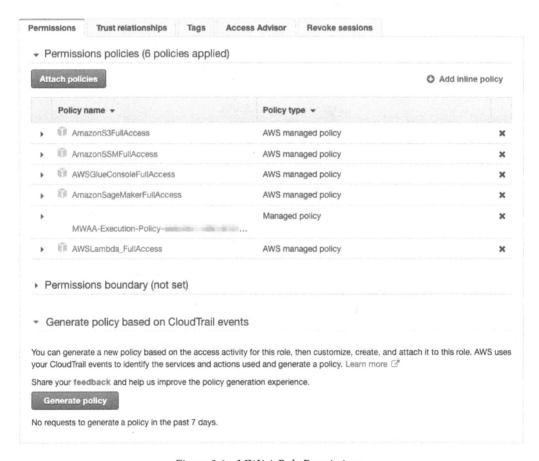

Figure 8.6 – MWAA Role Permissions

26. As you can see from *Figure 8.6*, we have added the necessary access to the various AWS services that the Airflow DAG will be leveraging.

> **Note**
>
> Providing full access to the necessary AWS services is not recommended in a production environment. We leverage these policies here for simplicity, within the context of our example.

27. After approximately 20 to 30 minutes, the MWAA environment should be **Available**. *Figure 8.7* shows what the **Airflow environments** screen should look like:

Figure 8.7 – Airflow environments

As you can see from *Figure 8.7*, the MWAA environment has been deployed and is available for us. Now we have all the prerequisites as well as an Airflow platform. In the next chapter, we will use this MWAA environment to create an automated, data-centric ML process.

Summary

In this chapter, we introduced a new approach to automating the ML workflow on AWS, namely, the data-centric approach. To orchestrate this data-centric workflow, we introduced a platform, typically used by data engineering teams, called Apache Airflow, and showed how to build such an environment using Amazon MWAA.

In the next chapter, we will see how to continue using the environment we've just created and create a DAG to automate the ML process for creating the *Age Calculator* model.

9

Building the ML Workflow Using Amazon Managed Workflows for Apache Airflow

In previous iterations of the *Age Calculator* example, we learned how applying a source code-centric methodology for ML workflow automation has been accomplished through cross-functional collaboration between the ML practitioner and developer teams. In *Chapter 8, Automating the Machine Learning Process Using Apache Airflow*, we explained how data engineering teams can use Amazon's MWAA to create the platform where the ML practitioner can automate the ML workflow as an Airflow DAG.

So, to build a successful data-centric ML workflow, we need to apply the same methodology to create an agile, cross-functional collaboration between the ML practitioner and data engineering teams. Therefore, in this chapter, we are going to continue where we left off in *Chapter 8*, *Automating the Machine Learning Process Using Apache Airflow*. In the previous chapter, we used the AWS CDK to construct the MWAA prerequisites, namely a Lambda Function to analyze the results from an ML model evaluation, a Glue Catalog to store our training data, a Glue Job to merge the new training data with the data already stored in the catalog, and a Codebuild project to sync an Airflow DAG with the MWAA environment. Along with these CDK artifacts, we also created an MWAA environment that will execute the data-centric workflow.

Thus, the primary motivation for this chapter is to highlight just how both the data engineering and ML practitioner teams can construct, execute, and manage the automated ML process on Apache Airflow by building and executing the Airflow DAG that's responsible for this data-centric workflow. By the end of the chapter, you will know how adding new training data will trigger the automated, end-to-end ML process and be able to generate a production-grade *Age Calculator* model.

To accomplish this, we will cover the following topics:

- Developing the data-centric workflow
- Creating synthetic Abalone survey data
- Executing the data-centric workflow

Technical requirements

For this chapter, you will need the following:

- A web browser (for the best experience, it is recommended that you use either Chrome or Firefox.)
- Access to the AWS account that you used in *Chapter 8*, *Automating the Machine Learning Process Using Apache Airflow*.
- Access to the Cloud9 development environment we used in *Chapter 8*, *Automating the Machine Learning Process Using Apache Airflow*.

- A reference to the usage limits of the AWS Free Tier to avoid exceeding unnecessary costs.

- The source code examples for this chapter, which are provided in this book's GitHub repository: `https://github.com/PacktPublishing/Automated-Machine-Learning-on-AWS/tree/main/Chapter09`). The code examples should already be available in the Cloud9 development environment. If not, please refer to the *Developing the application artifacts* section of *Chapter 4, Continuous Integration and Continuous Delivery (CI/CD) for Machine Learning*.

Developing the data-centric workflow

In *Chapter 8, Automating the Machine Learning Process Using Apache Airflow*, we created the environment components that are required to execute the data-centric ML workflow. Now, we can start developing it. The following diagram shows what this workflow development process looks like:

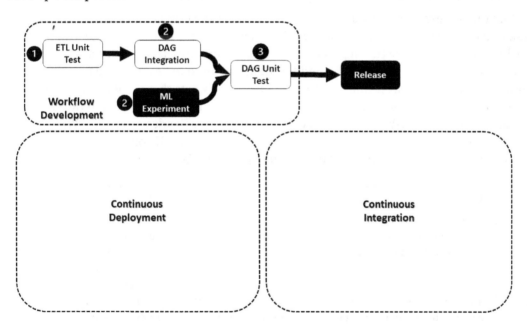

Figure 9.1 – Workflow development process

As you can see, the data engineering teams must develop two primary artifacts that make up the overall process, as follows:

- The unit tested data ETL artifacts
- The unit tested Airflow DAG

Once the data engineering team has created and tested the ETL artifacts that are responsible for merging and preparing the training data, they can combine them with the ML model artifacts to create the Airflow DAG, which represents the data-centric workflow. Upon unit testing this Airflow DAG, to ensure that both the data transformation code and the ML model code successfully integrate, the resultant workflow can be released to production.

Let's start building the ETL artifacts from the perspective of the data engineering team.

Building and unit testing the data ETL artifacts

Within the context of the overall data-centric workflow, the primary goal behind the ETL task is to merge any new data with the existing data so that the resulting dataset can be further split into separate training, validation, and testing datasets. However, as the data engineering team builds the code behind this task, they need to bear in mind that it's not always possible to pre-determine the exact amount of new data that needs to be merged. So, in this section, we will create the code artifacts for the ETL task and ensure the task is scalable by using an AWS Glue Job to execute the task as a Spark script. The AWS Glue Job that will be used to execute the ETL task was created as a CDK construct in *Chapter 8, Automating the Machine Learning Process Using Apache Airflow.*

> **Tip**
>
> To help the data engineering team create and unit test this Spark script, AWS has provided a Docker container that's bundled with the necessary libraries to construct and test Glue ETL Jobs. AWS has published this information in a blog post entitled Developing AWS Glue ETL jobs locally using a container (`https://aws.amazon.com/blogs/big-data/developing-aws-glue-etl-jobs-locally-using-a-container/`). Should you use the solution referenced within the blog post, it is recommended that you install version **0.24.1** of the pandas library. This version of pandas is required to copy a CSV file directly to S3.

So, to start building the ETL Job, we will create a Python script called `preprocess.py`. This script will read from the Glue Catalog, which contains the updated Abalone data and merge this with the original Abalone dataset and provide the overall feature transformations that are needed by the ML model.

> **Note**
>
> Since the core focus of this book is not on how to construct a Spark script, the basis for the `preprocess.py` file comes from the AWS SageMaker GitHub repository (`https://github.com/aws/amazon-sagemaker-examples/tree/master/advanced_functionality/inference_pipeline_sparkml_xgboost_abalone`). This example is licensed under the **Apache 2.0** License. We will build upon this example and customize it for our use case.

To create the ETL script, we will continue using the Cloud9 environment. Follow these steps:

1. Using the navigation panel of the Cloud9 environment, navigate to the `abalone-data-pipeline` folder.

    > **Note**
    >
    > We created the `abalone-data-pipeline` folder in *Chapter 8*, *Automating the Machine Learning Process Using Apache Airflow*.

2. Within the `abalone-data-pipeline` folder, expand the `artifacts` folder, and then expand the `airflow` folder. Right-click on the `scripts` folder and select the **New File** option. Create a file called `preprocess.py` and open it for editing.

3. Within the `preprocess.py` file, add the following code to import the necessary PySpark and AWS Glue libraries:

```
import sys
import os
import boto3
import pyspark
import pandas as pd
from functools import reduce
from pyspark.sql import SparkSession, DataFrame
from pyspark.ml import Pipeline
from pyspark.sql.types import StructField, StructType,
```

```
StringType, DoubleType
from pyspark.ml.feature import StringIndexer,
VectorIndexer, OneHotEncoder, VectorAssembler
from pyspark.sql.functions import *
from awsglue.job import Job
from awsglue.transforms import *
from awsglue.context import GlueContext
from pyspark.context import SparkContext
from awsglue.utils import getResolvedOptions
from awsglue.dynamicframe import DynamicFrame
from awsglue.utils import getResolvedOptions
```

4. Next, we will build some utility functions to help process the data. The first utility function is called `csv_line()`, whereby we supply a line of text data from a Spark **Resilient Distributed Dataset** (**RDD**) and create a comma-delimited string. This string will eventually be written to a CSV file on S3:

```
def csv_line(data):
    r = ','.join(str(d) for d in data[1])
    return str(data[0]) + "," + r
```

5. The next function that we will create is called `toS3()`. This function extracts the relevant features from the dataset, including the target feature, calls the `csv_line()` function to create a comma-delimited string for each line, converts the dataset into a pandas DataFrame, and writes the DataFrame to S3:

```
def toS3(df, path):
    rdd = df.rdd.map(lambda x: (x.rings, x.features))
    rdd_lines = rdd.map(csv_line)
    spark_df = rdd_lines.map(lambda x: str(x)).map(lambda
s: s.split(",")).toDF()
    pd_df = spark_df.toPandas()
    pd_df = pd_df.drop(columns=["_3"])
    pd_df.to_csv(f"s3://{path}", header=False,
index=False)
```

> **Note**
> Working with Spark DataFrames allows us to overcome the memory
> limitations of pandas DataFrames by distributing the dataset across multiple
> Spark nodes. However, when a Spark DataFrame is written to disk, it creates
> multiple *part* files, depending on the number of RDD partitions. To create a
> single CSV file, we must convert the Spark DataFrame into a single pandas
> DataFrame, thus writing the dataset to a single file. Using this technique may
> not scale if the single pandas DataFrame exceeds certain memory limitations.
> However, since the example dataset is not large, we can use pandas to create a
> single file.

6. Finally, we must create the `main()` function. Using the following code
    snippet, we can initialize the `spark` and `glueContext()` classes to wrap the
    `SparkContext` object:

```
. . .
def main():
    glueContext = GlueContext(SparkContext.getOrCreate())
    spark = SparkSession.builder.
appName("PySparkAbalone").getOrCreate()
    spark.sparkContext._jsc.hadoopConfiguration().
set("mapred.output.committer.class", "org.apache.hadoop.
mapred.FileOutputCommitter")
. . .
```

> **Note**
> Since AWS Glue is essentially a serverless Spark Processing Job,
> `SparkContext` represents the connection to the serverless Spark
> cluster, which is created and managed in the background by AWS. For
> more information on the `SparkContext` class, please refer to the Spark
> documentation (`https://spark.apache.org/docs/latest/`
> `api/java/org/apache/spark/SparkContext.html`).

7. Since we will be passing the `preprocess.py` file as a script, along with the
    command argument to AWS Glue, the following code snippet shows how we can
    declare the `args` variable using the `getResolvedOptions()` function. This is a
    utility function that's provided by AWS Glue to access the command arguments that
    are passed with the `preprocess.py` script:

```
. . .
    args = getResolvedOptions(sys.argv, ["GLUE_CATALOG",
```

```
"S3_BUCKET", "S3_INPUT_KEY_PREFIX", "S3_OUTPUT_KEY_
PREFIX"])
...
```

8. To read in the Abalone data as a Spark DataFrame, we must supply the appropriate schema for each of the column data types. In the following code snippet, we're declaring the `schema` variable, which contains the type or structure of the data that's found in each column of the dataset:

```
...
    schema = StructType(
        [
            StructField("sex", StringType(), True),
            StructField("length", DoubleType(), True),
            StructField("diameter", DoubleType(), True),
            StructField("height", DoubleType(), True),
            StructField("whole_weight", DoubleType(),
True),
            StructField("shucked_weight", DoubleType(),
True),
            StructField("viscera_weight", DoubleType(),
True),
            StructField("shell_weight", DoubleType(),
True),
            StructField("rings", DoubleType(), True)
        ]
    )
...
```

9. Next, we must write the following code to merge the new data from the Glue Catalog, along with the original Abalone dataset, to create a DataFrame called `distinct_df`. This DataFrame is strict in the sense that any duplicate rows are removed after the merge process:

```
...
    columns = ["sex", "length", "diameter", "height",
"whole_weight", "shucked_weight", "viscera_weight",
"shell_weight", "rings"]
    new = glueContext.create_dynamic_frame_from_
catalog(database=args["GLUE_CATALOG"], table_name="new",
transformation_ctx="new")
```

```
    new_df = new.toDF()
    new_df = new_df.toDF(*columns)
    raw_df = spark.read.csv(("s3://{}".format(os.path.
join(args["S3_BUCKET"], args["S3_INPUT_KEY_PREFIX"])))),
header=False, schema=schema)
    merged_df = reduce(DataFrame.unionAll, [raw_df, new_
df])
    distinct_df = merged_df.distinct()
    ...
```

10. Now that we have a unique DataFrame, we can set up the ETL pipeline and start transforming the dataset to prepare it for ML training. As shown in the following code snippet, the first part of the ETL process is to index the sex column as a training feature using the StringIndexer() class. Once the sex feature has been indexed, we can categorically encode the feature, thus creating vectors for each gender, by using the OneHotEncoder() class:

```
    ...
    sex_indexer = StringIndexer(inputCol="sex",
outputCol="indexed_sex")
    sex_encoder = OneHotEncoder(inputCol="indexed_sex",
outputCol="sex_vec")
    ...
```

11. The output of OneHotEncoder is a new set of columns, called sex_vec, that represent each gender. The next step is to use the VectorAssembler() class to merge the sex_vec columns with the original columns of the dataset. As shown in the following code snippet, here, we must instantiate VectorAssembler and define the assembler variable:

```
    ...
    assembler = VectorAssembler(
        inputCols=[
            "sex_vec",
            "length",
            "diameter",
            "height",
            "whole_weight",
            "shucked_weight",
            "viscera_weight",
```

```
        "shell_weight"
    ],
    outputCol="features"
)
...
```

12. As shown in the following code snippet, by combining `sex_indexer`, `sex_encoder`, and `assembler` into a `Pipeline`, and then fitting it onto the `distinct_df` DataFrame, we have a preprocessed `transformed_df` DataFrame, ready for model training:

```
...
    pipeline = Pipeline(stages=[sex_indexer, sex_encoder,
assembler])
    model = pipeline.fit(distinct_df)
    transformed_df = model.transform(merged_df)
...
```

13. The final step is to split the data into the relative training, validation, and testing datasets. As shown in the following code snippet, we must call the `toS3()` function to write these datasets to the data bucket as CSV files:

```
    (train_df, validation_df, test_df) = transformed_
df.randomSplit([0.8, 0.15, 0.05])
    toS3(train_df, os.path.join(args["S3_BUCKET"],
args["S3_OUTPUT_KEY_PREFIX"], "training/training.csv"))
    toS3(validation_df, os.path.join(args["S3_BUCKET"],
args["S3_OUTPUT_KEY_PREFIX"], "training/validation.csv"))
    toS3(test_df, os.path.join(args["S3_BUCKET"],
args["S3_OUTPUT_KEY_PREFIX"], "testing/testing.csv"))
...
```

14. With that, the main program is created to execute the data preprocessing task:

```
...
if __name__ == "__main__":
    main()
...
```

This completes the ETL script for the Glue Job. After unit testing the script, the data engineer can start creating the data processing workflow itself, in the form of an Airflow DAG. Let's go ahead and start building this DAG.

Building the Airflow DAG

Now, the data engineer must create the overall workflow for Airflow to execute, in the form of a DAG. You will recall that an Airflow DAG is a set of consecutive tasks, or operations, that are performed by the Airflow workers. To streamline the process of creating these consecutive operations, Airflow provides several prepackaged **operators**. These operators essentially encompass the logic for each task within the workflow. Since we are offloading the actual execution of these operations to AWS services, such as Glue and SageMaker, AWS provides several pre-built operators (`https://airflow.apache.org/docs/apache-airflow-providers-amazon/stable/operators/index.html`) for these and many other services.

However, using these AWS provider operators requires the data engineer or the ML practitioner to fully understand the relevant task operator, and thus how the AWS service executes the task. To simplify the DAG creation process, we will mostly use the standard `PythonOperator()` class (`https://airflow.apache.org/docs/apache-airflow/stable/_api/airflow/operators/python/index.html?highlight=pythonoperator#airflow.operators.python.PythonOperator`) to call the SageMaker service. This means that the data engineer can copy and paste the SageMaker SDK code from the ML experiment notebook into the workflow DAG. Doing this makes it easier for both the ML practitioner and data engineer to integrate the ML process into the data workflow. As you will see, using the `PythonOperator()` class within the DAG allows for further customizations to be made that the AWS provider operators don't provide.

> **Note**
>
> The AWS team provides numerous examples that showcase how to leverage the AWS provider operators for SageMaker (`https://sagemaker.readthedocs.io/en/stable/workflows/airflow/index.html`). However, since we will be using `PythonOperator()` to construct the SageMaker service calls, we will be basing our solution on another AWS example (`https://github.com/aws/amazon-sagemaker-examples/blob/master/sagemaker-experiments/track-an-airflow-workflow/track-an-airflow-workflow.ipynb`). This example is provided under the **Apache 2.0** License. We will be building on this example to make our DAG resemble the ML workflow we configured in *Chapter 7, Building the ML Workflow Using AWS Step Functions*. You can review the ML workflow by referencing *Figure 7.1* in the *Creating the ML workflow* section.

To start building the Airflow DAG, follow these steps:

1. Within your Cloud9 workspace, right-click on the `dags` folder and select the **New Folder** option to create a folder called `model`.

2. To define the Lambda Function handler, we can reuse the code we created in *Chapter 7, Building the ML Workflow Using AWS Step Functions*. Run the following command to copy the `model_training.py` file from the already cloned companion GitHub repository:

    ```
    $ cp ~/environment/src/Chapter09/Files/airflow/dags/
    model/model_training.py ~/environment/abalone-data-
    pipeline/artifacts/airflow/dags/model/
    ```

    > **Note**
    >
    > If you need to refamiliarize yourself with the `model_training.py` file, you can review the code in the *Creating the ML workflow* section of *Chapter 7, Building the ML Workflow Using AWS Step Functions*.

3. Next, right-click on the `dags` folder and select the **New File** option, creating a file called `abalone_data_pipeline.py`.

4. Double-click on the file for editing and add the following code to import the base Python libraries:

    ```
    import boto3
    import json
    from datetime import timedelta
    ```

5. Next, we must add the following SageMaker SDK libraries:

    ```
    import sagemaker
    from sagemaker.tensorflow import TensorFlow
    from sagemaker.tensorflow.serving import Model
    from sagemaker.processing import ProcessingInput,
    ProcessingOutput, Processor
    from sagemaker.model_monitor import DataCaptureConfig
    ```

    > **Note**
    >
    > These are the same SageMaker SDK libraries that the ML practitioner uses for the ML experiment notebook.

6. Now, using the following code, we can import the AWS provider operators, as well as the Airflow provider operators:

```
import airflow
from airflow import DAG
from airflow.operators.python_operator import
PythonOperator
from airflow.providers.amazon.aws.operators.glue import
AwsGlueJobOperator
from airflow.providers.amazon.aws.operators.glue_crawler
import AwsGlueCrawlerOperator
from airflow.providers.amazon.aws.hooks.lambda_function
import AwsLambdaHook
from airflow.operators.python_operator import
BranchPythonOperator
from airflow.operators.dummy import DummyOperator
```

7. Next, we must use the following code to define our global variables, as well as get the stored parameters that we defined in the CDK application:

```
sagemaker_seesion = sagemaker.Session()
region_name = sagemaker_seesion.boto_region_name
model_name = "abalone"
data_prefix = "abalone_data"
data_bucket = f"""{boto3.client("ssm", region_
name=region_name).get_parameter(Name="AirflowDataBucket")
["Parameter"]["Value"]}"""
glue_job_name = f"""{boto3.client("ssm", region_
name=region_name).get_parameter(Name="GlueJob")
["Parameter"]["Value"]}"""
crawler_name = f"""{boto3.client("ssm", region_
name=region_name).get_parameter(Name="GlueCrawler")
["Parameter"]["Value"]}"""
sagemaker_role = f"""{boto3.client("ssm", region_
name=region_name).get_parameter(Name="SageMakerRoleARN")
["Parameter"]["Value"]}"""
lambda_function = f"""{boto3.
client("ssm", region_name=region_name).get_
parameter(Name="AnalyzeResultsLambda")["Parameter"]
["Value"]}"""
container_image = f"763104351884.dkr.ecr.{region_name}.
```

```
amazonaws.com/tensorflow-training:2.5.0-cpu-py37-
ubuntu18.04-v1.0"
training_input = f"s3://{data_bucket}/{data_prefix}/
training"
testing_input = f"s3://{data_bucket}/{data_prefix}/
testing"
data_capture = f"s3://{data_bucket}/endpoint-data-
capture"
```

> **Note**
>
> As we saw in *Chapter 7, Building the ML Workflow Using AWS Step Functions*, SageMaker Processing Jobs don't provide TensorFlow containers. Therefore, we must leverage the Deep Learning TensorFlow container and reference it using the `container_image` variable.

8. The final variable we must define is `default_args` for the Airflow DAG. In the following code, we have specified some of the defaults that the Airflow scheduler requires to execute the DAG:

```
default_args = {
    "owner": "airflow",
    "depends_on_past": False,
    "start_date": airflow.utils.dates.days_ago(1),
    "retries": 0,
    "retry_delay": timedelta(minutes=2)
}
```

9. Since we are using the `PythonOperator()` class to interface with the SageMaker service, we must define multiple functions that encapsulate the logic of the service call. As we mentioned previously, these functions can be cut and pasted from the ML experiment notebook. For example, the following code creates the `training()` function, which utilizes the SageMaker SDK to create a `TensorFlow()` estimator, and calls the `fit()` method to train the model as a SageMaker Training Job:

```
def training(data, **kwargs):
    estimator = TensorFlow(
        base_job_name=model_name,
        entry_point="/usr/local/airflow/dags/model/model_
training.py",
```

```
        role=sagemaker_role,
        framework_version="2.4",
        py_version="py37",
        hyperparameters={"epochs": 200, "batch-size": 8},
        script_mode=True,
        instance_count=1,
        instance_type="ml.m5.xlarge",
    )
    estimator.fit(data)
    kwargs["ti"].xcom_push(
        key="TrainingJobName",
        value=str(estimator.latest_training_job.name)
    )
```

10. The next Python function, called `evaluation()`, executes a SageMaker Processing Job to execute the `evaluate.py` file that we created in *Chapter 8, Automating the Machine Learning Process Using Apache Airflow*, and evaluate the trained model's inference on the test dataset. The following code snippet shows how the `evaluation()` function is instantiated – that is, by specifying the name of the SageMaker Training Job that was defined in *Step 9* to instantiate it as a TensorFlow estimator so that we can get the location of the trained model:

```
...
def evaluation(ds, **kwargs):
    training_job_name = kwargs["ti"].xcom_
pull(key="TrainingJobName")
    estimator = TensorFlow.attach(training_job_name)
    model_data = estimator.model_data,
...
```

11. As part of the `evaluation()` function, we must also define the processor variable to initialize an instance of the SageMaker `Processor` class. The following code snippet shows how we must provide the necessary parameters to execute the Processing Job, namely the Processing Job name, the location of the processing container image, the processing script to execute, the SageMaker IAM role, and the type of compute resources to use for the Processing Job:

```
...
    processor = Processor(
        base_job_name=f"{model_name}-evaluation",
```

```
        image_uri=container_image,
        entrypoint=[
            "python3",
            "/opt/ml/processing/input/code/evaluate.py"
        ],
        instance_count=1,
        instance_type="ml.m5.xlarge",
        role=sagemaker_role,
        max_runtime_in_seconds=1200
    )
...
```

12. The following code snippet shows how the `processor.run()` method is called to execute the Processing Job we defined in *Step 11*. To run the Processing Job, we must supply the S3 location of the test dataset (`testing_input`), the S3 location of the trained ML model (`model_data`), and the S3 location of the `evaluate.py` script:

```
...
    processor.run(
        inputs=[
            ProcessingInput(
                source=testing_input,
                destination="/opt/ml/processing/testing",
                input_name="input"
            ),
            ProcessingInput(
                source=model_data[0],
                destination="/opt/ml/processing/model",
                input_name="model"
            ),
            ProcessingInput(
                source="s3://{}/airflow/scripts/evaluate.
py".format(data_bucket),
                destination="/opt/ml/processing/input/
code",
                input_name="code"
```

```
            )
        ],
    ...
```

13. Along with the defining `inputs` in *Step 12*, the following code snippet shows how to define the S3 location for the Processing Job results as an `output` parameter:

```
    ...
        outputs=[
            ProcessingOutput(
                source="/opt/ml/processing/evaluation",
                destination="s3://{}/{}/evaluation".
    format(data_bucket, data_prefix),
                output_name="evaluation"
            )
        ]
    )
```

14. Now that we have functions to train and evaluate the ML model, we must define a function to deploy the trained model as a SageMaker Hosted Endpoint by using the `deploy()` method on the trained TensorFlow estimator:

```
def deploy_model(ds, **kwargs):
    training_job_name = kwargs["ti"].xcom_
pull(key="TrainingJobName")
    estimator = TensorFlow.attach(training_job_name)
    model = Model(
        model_data=estimator.model_data,
        role=sagemaker_role,
        framework_version="2.4",
        sagemaker_session=sagemaker.Session()
    )
    model.deploy(
        initial_instance_count=2,
        instance_type="ml.m5.large",
        data_capture_config=DataCaptureConfig(
            enable_capture=True,
            sampling_percentage=100,
            destination_s3_uri=data_capture
```

```
        )
    )
```

15. Previously, as part of the CDK application, we defined a Lambda function to calculate the **RMSE** evaluation metric for the trained model. In the following code, we are leveraging the AWS provider operator to invoke this Lambda function:

```
def get_results(ds, **kwargs):
    hook = AwsLambdaHook(
        function_name=lambda_function,
        aws_conn_id="aws_default",
        invocation_type="RequestResponse",
        log_type="None",
        qualifier="$LATEST",
        config=None
    )
    request = hook.invoke_lambda(
        payload=json.dumps(
            {
                "Bucket": data_bucket,
                "Key": f"{data_prefix}/evaluation/
evaluation.json"
            }
        )
    )
    response = json.loads(request["Payload"].read().
decode())
    kwargs["ti"].xcom_push(
        key="Results",
        value=response["Result"]
    )
```

16. The last function we must create will take the RMSE score and compare it to the evaluation threshold to determine whether the trained model is considered production-grade. If the evaluation is approved, the model will be deployed as a SageMaker Hosted Endpoint. Alternatively, if the model is above the predefined threshold, the workflow will be categorized as rejected:

```
def branch(ds, **kwargs):
```

Developing the data-centric workflow   293

```
result = kwargs["ti"].xcom_pull(key="Results")
if result > 3.1:
    return "rejected"
else:
    return "approved"
```

> **Note**
>
> To ensure that this workflow example completes successfully and deploys the trained model, we must set the evaluation threshold higher than the threshold we used in *Chapter 7, Building the ML Workflow Using AWS Step Functions*. After successfully testing the Airflow DAG, you can set the threshold lower to mimic a more realistic ML model evaluation.

17. Now that we have created the processing logic for each step of the workflow, we can use the following code to define the DAG itself. Here, we are using the DAG() class to initialize the DAG, provide the name of the workflow and the default arguments, and schedule the DAG to automatically execute every night at midnight:

```
with DAG(
    dag_id=f"{model_name}-data-workflow",
    default_args=default_args,
    schedule_interval="@daily",
    concurrency=1,
    max_active_runs=1,
) as dag:
```

18. The first step that the DAG executes is crawler_task. Here, the Airflow Scheduler calls the AWS Glue Crawler to read the new data, infer the data schema, and append the data to the Glue Catalog. In the following code, we are defining the task using the AWS-provided AwsGlueCrawlerOperator():

```
crawler_task = AwsGlueCrawlerOperator(
    task_id="crawl_data",
    config={"Name": crawler_name}
)
```

19. The second step of the workflow is called `etl_task`. In this task, we call the AWS-provided `AwsGlueJobOperator()` to run the Glue Job we defined in the CDK application. You will recall that this Job merges the initial Abalone dataset with the new data from the Glue Catalog, and then preprocesses it to create the training, validation, and test datasets:

```
etl_task = AwsGlueJobOperator(
    task_id="preprocess_data",
    job_name=glue_job_name
)
```

20. Now that the dataset has been prepared and stored in the data bucket, we can use the `PythonOperator()` class to call our `training()` function. This task supplies the location of the preprocessed training data and calls SageMaker to run a Training Job using the TensorFlow estimator:

```
training_task = PythonOperator(
    task_id="training",
    python_callable=training,
    op_args=[training_input],
    provide_context=True,
    dag=dag
)
```

21. The next task in the workflow is `evaluation_task`. Here, we're using `PythonOperator()` to call the `evaluation()` function, whereby we instruct SageMaker to execute a Processing Job and test the trained model against the testing dataset:

```
evaluation_task = PythonOperator(
    task_id="evaluate_model",
    python_callable=evaluation,
    provide_context=True,
    dag=dag
)
```

> **Note**
>
> Note that in the evaluation() function, we use Airflow **cross-communications**, or **Xcoms** (https://airflow.apache.org/docs/apache-airflow/stable/concepts/xcoms.html), to pass the name of the SageMaker Training Job between tasks. This is one of the primary reasons we leverage the PythonOperator() class instead of the AWS-provided operators for SageMaker.

22. The next task is the Lambda function that determines the model evaluation results from the test dataset. analyze_results_task uses PythonOperator() to call the get_results() Python function. You will recall that this Python function causes the AnalyzeResults Lambda function to return the RMSE score:

```
analyze_results_task = PythonOperator(
    task_id="analyze_results",
    python_callable=get_results,
    provide_context=True,
    dag=dag
)
```

23. Based on the returned RMSE results, the next task within the workflow is to determine whether the model is ready for production. Here, we're using the BranchPythonOperator() class to call the branch() Python function and evaluate the returned results against the pre-determined threshold:

```
check_threshold_task = BranchPythonOperator(
    task_id="check_threshold",
    python_callable=branch,
    provide_context=True,
    dag=dag
)
```

24. Should the model evaluation result be lower than the threshold value, the workflow will move on to deployment_task. This task calls the deploy_model() Python function to create a SageMaker Hosted Endpoint:

```
deployment_task = PythonOperator(
    task_id="deploy_model",
    python_callable=deploy_model,
    provide_context=True,
```

```
        dag=dag
    )
```

25. Finally, we must create the placeholder tasks by using DummyOperator() to create placeholders for the start, end, rejected, and approved states within the workflow:

```
start_task = DummyOperator(
    task_id="start",
    dag=dag
)
end_task = DummyOperator(
    task_id="end",
    dag=dag
)
rejected_task = DummyOperator(
    task_id="rejected",
    dag=dag
)
approved_task = DummyOperator(
    task_id="approved",
    dag=dag
)
```

26. Now that the various tasks of the workflow have been defined, we must create the overall flow of the DAG. In the following code, we're defining the dependencies between each of the specific tasks:

```
start_task >> crawler_task >> etl_task >> training_
task >> evaluation_task >> analyze_results_task >> check_
threshold_task >> [rejected_task, approved_task]
approved_task >> deployment_task >> end_task
rejected_task >> end_task
```

27. The workflow DAG is now complete. Now, we must save the file and run the following commands in the Cloud9 Terminal window to check the code into the repository:

```
$ git add -A
$ git commit -m "Initial commit of workflow DAG"
$ git push
```

> **Note**
>
> Before pushing the DAG to the CodeCommit repository, the data engineer may want to perform local unit tests to ensure that the DAG is fully functional before it is imported by MWAA. AWS provides a command-line interface utility called **aws-mwaa-local-runner** (https://github.com/aws/aws-mwaa-local-runner) that reproduces an MWAA environment locally using a Docker container. By using this utility, the data engineer can not only unit test the DAG, but also verify that the Python dependencies will work on MWAA (https://docs.aws.amazon.com/mwaa/latest/userguide/working-dags-dependencies.html).

Now that we've created the workflow DAG and its associated artifacts we must commit the changes to the CodeCommit repository. This will cause the build to deploy these files to the data bucket. Once there, and after about 5 minutes, the MWAA scheduler will import the DAG. You can now view the DAG in the MWAA web UI. The following steps will walk you through how to access the MWAA web UI:

1. Open the MWAA console (https://console.aws.amazon.com/mwaa/home) and select your MWAA environment, called **MyAirflowEnvironment**.

2. Click the **Open Airflow UI** link to open the MWAA web UI.

3. Once the UI has opened in a new browser tab, you should eventually see the **abalone-data-pipeline** DAG. The following screenshot shows an example of the newly imported **abalone-data-workflow** DAG in the web UI:

DAGs

Figure 9.2 – abalone-data-workflow DAG

4.   Click on the DAG to open it.

> **Note**
>
> Do not enable the DAG just yet since we haven't supplied any new data for the
> workflow to successfully execute.

5.   Click on the **Graph View** link to view the DAG as a graph. The following diagram
      shows the overall workflow as a graph:

Figure 9.3 – Workflow graph

Note that this data-centric workflow is similar to the ML workflow we created in *Chapter
7, Building the ML Workflow Using AWS Step Functions*, except for the Glue Crawler and
scalable Glue ETL Job. However, before we can see the workflow in action, we need to
simulate the process of adding new Abalone survey data. Let's get started.

Creating synthetic Abalone survey data

In the previous section, we created the two primary artifacts – from the perspective of
the data engineering team – that are required to successfully implement the data-centric
workflow, with the first being the ETL artifacts that merge the raw Abalone data with
new data to create the training, validation, and test datasets. We also integrated these ETL
artifacts into a data-centric workflow, in the form of an Airflow DAG artifact, to automate
the ML process whereby we can train, evaluate, and deploy a production-grade *Age
Calculator* model.

As you may recall from the *Using Airflow to process the Abalone dataset* section of *Chapter
8, Automating the Machine Learning Process Using Apache Airflow*, we established the
context for the data-centric workflow by expanding the ACME Fishing Logistics use case
to address the need to add updated Abalone survey data.

So, before we can execute the data-centric workflow, we must address the next step. The
following diagram illustrates the next step we will be addressing in this section – that is,
simulating new Abalone survey data to further optimize the ML model:

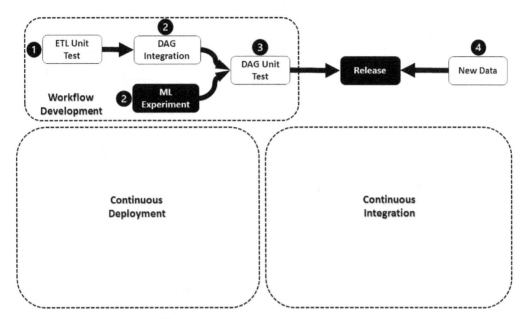

Figure 9.4 – Simulating new Abalone survey data

Since the Abalone Survey Company is a fictional entity in our example, we are going to have to somehow get new Abalone data; since there are no new sources for the data, we will have to synthesize some. Fortunately, the **Data to AI Group** at **MIT** (`https://dai.lids.mit.edu/`), has open sourced a project called **CTGAN** (`https://github.com/sdv-dev/CTGAN`) to help us synthesize new Abalone data.

> **Note**
>
> The CTGAN project is available under the MIT License (`https://github.com/sdv-dev/CTGAN/blob/master/LICENSE`).

CTGAN uses a deep learning-based **Synthetic Data Generator**, essentially a conditional generative adversarial network model, to learn from data and *predict* a new dataset. The following steps will walk you through how to leverage CTGAN to synthesize new Abalone data using the SageMaker Studio UI:

1. Open the SageMaker management console (`https://console.aws.amazon.com/sagemaker/home`), and then click on the **SageMaker Domain** option in the left-hand navigation panel.

2. Once the **SageMaker Domain** dashboard opens, click on the **Launch app** drop-down box and select the **Studio** option to open the Studio IDE.

> **Note**
>
> If you've been following this book, you should already have a configured SageMaker Studio environment. If not, please refer to the *Getting started with SageMaker Studio* section of *Chapter 2, Automating Machine Learning Model Development Using SageMaker Autopilot.*

3. From the **File** menu, click **New** and select **Notebook** to open a new Jupyter Notebook. This will create a new notebook called `Untitled.ipynb` in the `root` folder. Since we are using this notebook to synthesize new Abalone survey data, it can be created in any folder within your SageMaker Studio environment.

4. When prompted, select the **Python 3 (data Science)** kernel and click the **Select** button.

> **Note**
>
> It is recommended that you use an **ml.m5.4xlarge (16 vCPUs + 64 MB)** instance type. However, this will incur additional AWS usage costs. Additionally, an example Jupyter Notebook is available in this book's GitHub repository (`https://github.com/PacktPublishing/Automated-Machine-Learning-on-AWS/blob/main/Chapter09/Notebook/Simulating%20New%20Abalone%20Survey%20Data.ipynb`).

5. Once the kernel starts, use the following code in a code cell to install the CTGAN libraries:

```
%%capture
!pip install ctgan
```

6. Next, import the necessary Python libraries and global variables:

```
import io
import boto3
import warnings
import pandas as pd
from time import gmtime, strftime
warnings.filterwarnings("ignore")
s3 = boto3.client("s3")
model_name = "abalone"
column_names = [
    "sex",
```

```
        "length",
        "diameter",
        "height",
        "whole_weight",
        "shucked_weight",
        "viscera_weight",
        "shell_weight",
        "rings"
    ]
```

7. In the next code cell, add the following code to open the original (or *raw*) Abalone dataset that was uploaded to S3 when we deployed the CDK application, as well as define the name of the new Abalone data file. The new data file will contain the current date and time appended to the filename, making it a unique survey:

```
data_bucket = f"""{boto3.client("ssm").get_
parameter(Name="AirflowDataBucket")["Parameter"]
["Value"]}"""
raw_data_key = f"{model_name}_data/raw/abalone.data"
new_data_key = f"{model_name}_data/new/abalone.
{strftime('%Y%m%d%H%M%S', gmtime())}"
s3_object = s3.get_object(Bucket=data_bucket, Key=raw_
data_key)
raw_df = pd.read_csv(io.BytesIO(s3_object["Body"].
read()), encoding="utf8", names=column_names)
```

8. Now, add the following code to fit a CTGAN model to the raw data, specifying the `sex` feature as a categorical value:

```
from ctgan import CTGANSynthesizer
ctgan = CTGANSynthesizer()
ctgan.fit(raw_df, ["sex"])
```

9. To generate the 100 samples of synthesized survey data, add the following code to a new code cell:

```
samples = ctgan.sample(100)
```

10. Now that we have new synthesized the Abalone data as the `samples` variable, we can use the following code to copy it to the S3 data bucket:

```
samples.to_csv(f"s3://{data_bucket}/{new_data_key}",
header=False, index=False)
```

With the new Abalone survey data synthesized and uploaded to S3, we can execute the data-centric workflow. We'll learn how to do this in the next section.

Executing the data-centric workflow

In the previous section, we successfully generated new Abalone survey data. So, with this dataset now stored on S3, this section will walk you through how to execute and release the data-centric workflow to create a production-grade ML model that has been optimized on both the new, as well as the original, datasets.

As with the example in *Chapter 7, Building the ML Workflow Using AWS Step Functions*, we can consider this execution and any scheduled execution of the workflow as a release change. The following diagram shows an overview of the workflow execution that we defined within the Airflow DAG:

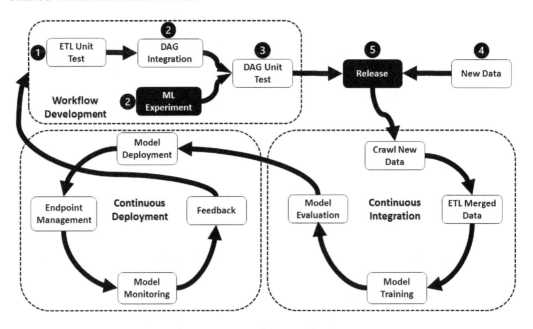

Figure 9.5 – Overview of the workflow's execution

As you can see, once we have new data and the schedule kicks off, the Airflow DAG will execute the CI phase of updating the Abalone dataset, training a new ML model, and evaluating the trained model's performance.

Once the model has been automatically approved as a production-grade model, it is deployed to production during the CD phase. The operations teams can then take ownership of the hosted model to manage and continuously monitor its production performance.

This CI/CD process will execute every night at midnight, based on the DAG schedule, to ensure that the production model is continuously optimized as it gets trained on new survey data.

To see this in action, perform the following steps to execute the workflow release:

1. Using the Airflow web UI, click the toggle button next to the **abalone-data-workflow** DAG to enable it.

2. Once the DAG has been enabled, the workflow will automatically start. Click on the DAG to view its execution.

3. Using either the **Tree View** or **Graph View** links, you can view each task of the DAG being executed. The following diagram shows what a completed workflow execution graph looks like:

Figure 9.6 – Completed workflow execution graph

4. Clicking on any of the tasks will allow you to view its task configuration and, more importantly, the log output from the worker nodes. Click on a task to open the **Task Instance** window, then click the **Log** button to open the worker logs:

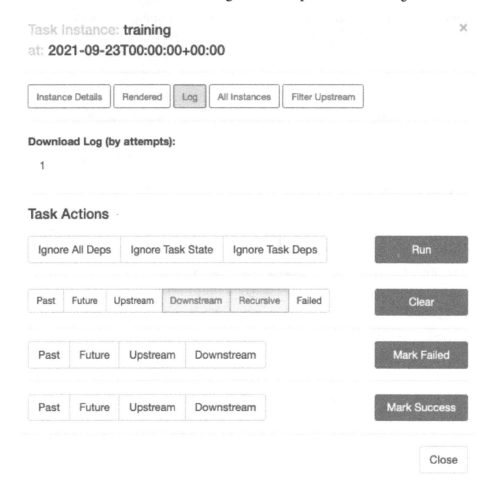

Figure 9.7 – Task Instance window

> **Note**
>
> Since the SageMaker tasks use `PythonOperator()`, the output from the logs shows a redirect of the SageMaker CloudWatch logs. This is another reason to make SageMaker execution calls using the SageMaker SDK and `PythonOperator()`, as opposed to the AWS-provided SageMaker operators, since these require you to view the log output in CloudWatch instead of the Airflow web UI.

5. To see the evaluation RMSE score, click on the **analyze_results** task instance and click on the **Log** button. Once the **Log** screen appears, click on the **XCom** button. As shown in the following screenshot, you can see **XCom** for the **Results** key. This key is available to the downstream **check_threshold** task to determine whether the model should be approved or rejected for production:

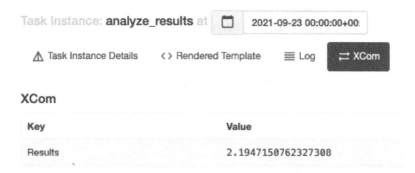

Figure 9.8 – Example RMSE

Using the preceding steps, we have created and executed a data-centric, automated ML workflow, which will also execute daily. Should new survey data be uploaded, the model will be trained on the original dataset, as well as the new survey data, hopefully making it more robust.

However, it is important to recognize that even though the workflow will be executed once a day, we have deployed infrastructure resources that will only be used during the scheduled execution. This means that the MWAA worker nodes are sitting idle when they're not being used, thus consuming AWS billable resources. To offset overspending for unused resources, we may want to review the minimum and maximum worker count for the MWAA environment and adjust it accordingly.

In the next section, you will learn how to limit the AWS costs for this example by deleting these various resources.

Cleanup

Follow these steps to remove the various resources we've deployed:

1. Open the SageMaker console (https://console.aws.amazon.com/ sagemaker/home) and, using the left-hand navigation menu, select **Inference**, and then the **Endpoints** option.

2. Delete any **Endpoints** by selecting the radio button next to each **Name** and clicking the **Actions** dropdown, then the **Delete** option.

3. Repeat this procedure for any **Endpoint configurations** and any **Models**.

4. Open the MWAA console (`https://console.aws.amazon.com/mwaa/home`) and select your environment. Click the **Actions** dropdown and select the **Delete** option to delete the MWAA environment.

> **Note**
> Wait for the MWAA environment to be deleted before proceeding with the next step.

5. Open the CloudFormation console (`https://console.aws.amazon.com/cloudformation/home`) and select **MWAA-VPC stack** by checking the radio button next to the stack. Once selected, click the **Delete** button.

6. Repeat the same procedure for the **abalone-data-pipeline** stack.

With that, we have successfully deleted the various AWS resources that we deployed both in this chapter and *Chapter 8, Automating the Machine Learning Process Using Apache Airflow*.

Summary

In this chapter, we expanded upon the data-centric approach that we introduced in the previous chapter to automate the ML workflow using Apache Airflow. To do this, we learned how to build the artifact that's responsible for merging the existing dataset with new data to optimize the Age Calculator model. We also learned how to use the CTGAN data generator to synthesize this new survey data. Once the new survey data was uploaded to S3, we learned how to build and then execute the Airflow DAG that's responsible for the data-centric workflow.

With this hands-on example, we learned how the platform, data engineering teams, and ML practitioners can work together to create a data-centric approach to ML automation. We also learned how AWS makes it easier to deploy, manage, and maintain an Apache Airflow environment with our implementation of an Amazon MWAA environment and, subsequently, use this environment to create a production-grade *Age Calculator* model.

In the next chapter, we will apply what we've learned in this and the previous chapter to learn how the data-centric approach can further augment the CI/CD methodology to include **continuous training** (**CT**), an additional phase of ML automation.

Section 5: Automating the End-to-End Production Application on AWS

This section will introduce you to the **Machine Learning Software Development Life Cycle (MLSDLC)** and how to implant the end-to-end process, with the ACME Fishing Logistics example. This section encompasses the various techniques learned in previous parts and shows how they fit into the MLSDLC. The various chapters within this part will introduce you to what the MLSDLC is and how it works in practice by highlighting the various roles of a cross-functional team and how team members implement the **CI** (**Continuous Integration**), **CD** (**Continuous Delivery**), and **CT** (**Continuous Training**) aspects of the production application.

This section comprises the following chapters:

- *Chapter 10, An Introduction to the Machine Learning Software Development Life Cycle (MLSDLC)*

- *Chapter 11, Continuous Integration, Deployment, and Training for the MLSDLC*

10

An Introduction to the Machine Learning Software Development Life Cycle (MLSDLC)

At this point in the book, we have reviewed multiple **Amazon Web Services** (**AWS**) technologies that can be used to automate the **machine learning** (**ML**) process, from automating ML experimentation with **Amazon SageMaker Autopilot** to automating model training and deployments with **AWS CodePipeline**, **AWS Step Functions**, and **Amazon Managed Workflows for Apache Airflow** (**MWAA**). We've also seen how various processes can be applied to the task of ML automation by reviewing both a source code-centric and a data-centric methodology to further optimize the ML process. Throughout the previous chapters, we've also seen how different teams within the organization can contribute to the overall success of the ML use case.

In this chapter, we're going to apply what we've already learned, and expand on the key factors that influence a successful execution of an automated, **end-to-end** (**E2E**) ML strategy or **ML software development life cycle** (**MLSDLC**), namely the following:

- Processes
- Technology
- People

We will expound on these factors by focusing on the various roles within a cross-functional, agile team, and the specific artifacts that each team contributes to creating a quality ML-based application, by covering the following topics:

- Introducing the MLSDLC
- Building the application platform
- Examining ML and data engineering roles
- Understanding the security lens

By the end of the chapter, you will have a fair idea of what the MLSDLC process encompasses, and how the process can be applied to the *Age Calculator* use case.

Technical requirements

To follow along with the examples in the chapter, you will need the following:

- A web browser (for the best experience, it is recommended that you use a Chrome or Firefox browser).
- Access to the **AWS Account** that you've been using through the book.
- Access to the **Cloud9** development environment we've been using thought the book.
- We will once again be working within the usage limits of the AWS Free Tier to avoid incurring unnecessary costs.
- The full source code for the application artifacts is provided in the companion GitHub repository for this chapter (`https://github.com/PacktPublishing/Automated-Machine-Learning-on-AWS/tree/main/Chapter10`).

Introducing the MLSDLC

The concept of a **systems development life cycle** (**SDLC**), or application development life cycle, has been around since the 1960s, whereby six individual processes are put in place to effectively **plan**, **design**, **build**, **test**, **deploy**, and **maintain** applications in production. While the individual phases of the process, as well as the mechanisms to implement these phases, have evolved over the years, the fundamental requirement to quickly and effectively deliver a working application into production hasn't. The following diagram shows a high-level overview of the six phases of the SDLC:

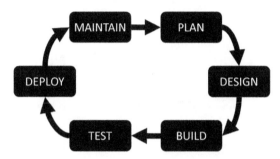

Figure 10.1 – Six phases of the SDLC

When looking closely at *Figure 10.1*, you should hopefully deduce a correlation with some of the processes we've encountered up until this point in the book. For example, we could assume that some of the potential activities performed during the plan and design phases of the SDLC might be similar to some of the activities that could be performed during the business use case phase of the CRISP-DM process. Recall that we reviewed the **Cross-Industry Standard Process for Data Mining** (**CRISP-DM**) process in *Chapter 1, Getting Started with Automated Machine Learning on AWS*.

Additionally, if we refer to the **continuous integration/continuous delivery** (**CI/CD**) process that was introduced in *Chapter 4, Continuous Integration and Continuous Delivery (CI/CD) for Machine Learning*, we could also deduce that there is a corresponding mapping of tasks between the **build**, **test**, **deploy**, and **maintain** phases of the SDLC. So, as we've worked through the various ways to implement a production-grade model for the *Age Calculator* use case, we have indirectly been creating an ML-focused SDLC— or MLSDLC.

The best way to demonstrate this assumption is to build out an example application (website) using the SDLC process and incorporate an ML use case (the *Age Calculator* model) to complete an MLSDLC process. So, in this chapter and the next, we are going to build an **ACME Fishing Logistics** website, and in doing so, not only highlight the MLSDLC process but also emphasize the critical factors that influence a successful MLSDLC implementation—namely, *the process* (CI/CD), *the people* (cross-functional team), and *the technology* (AWS services).

To set the stage, the following diagram shows the high-level architecture for the application platform we will be building:

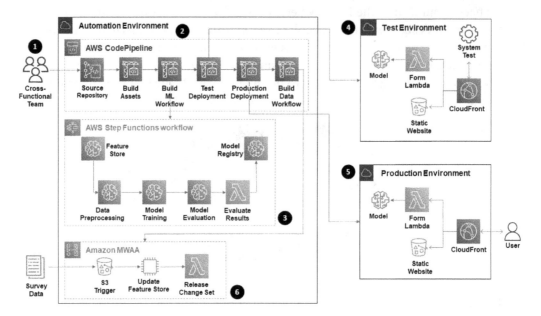

Figure 10.2 – ACME Fishing Logistics application platform

As you can see from *Figure 10.2*, the application platform uses several AWS services to deliver the final solution. We can group the components into six specific categories by following these next steps:

1. To build this solution, a cross-functional team creates various codified artifacts, encapsulating their contribution to the overall solution. We will be focusing on creating these artifacts in this chapter.

2. Once the artifacts are created and committed to the source code repository, the next component of the solution orchestrates the build and deployment of these artifacts as CodePipeline assets. For example, CodePipeline orchestrates building the Docker container image for model training. This process is similar to what we learned about in *Chapter 4, Continuous Integration and Continuous Delivery (CI/CD) for Machine Learning*.

3. After the pipeline assets have been built, CodePipeline then orchestrates the deployment of the automated ML workflow, in the form of a Step Functions state machine. Using the techniques we learned in *Chapter 6, Automating the Machine Learning Process using AWS Step Functions*, the workflow processes the training data, and then trains and evaluates a production-grade ML model. The final model is stored in the SageMaker model registry.

4. Accordingly, once we have a production-grade ML model, CodePipeline then orchestrates the deployment of a testing environment, utilizing the production-grade ML model stored in the SageMaker model registry. After the testing environment has been deployed, we then execute a system test on a pseudo-production version of the website application.

5. Once the test version of the website application passes the system tests, CodePipeline then deploys the production version of the website application. It's at this stage that we have implemented E2E automation of a production ML application, using a source code-centric approach.

6. To facilitate adding new abalone survey data, the final component that CodePipeline deploys is an Amazon MWAA environment, like the one we created in *Chapter 8, Automating the Machine Learning Process using Apache Airflow*. Therefore, by means of an Airflow **directed acyclic graph** (**DAG**) to update the training data, we will have incorporated a data-centric mechanism into the solution.

So, now that we have an idea of what we will be building in this chapter and the next chapter, in the next section, we kick the process off by starting with the planning, design, and build for the application platform.

Building the application platform

In *Chapter 1, Getting Started with Automated Machine Learning on AWS*, you were introduced to ACME Fishing Logistics, whose primary charter is to educate fishermen on how to determine whether abalone is old enough for breeding. To accomplish this task, ACME provides a website with the relevant information to guide fishermen in their abalone age-determination task. ACME also provides a contact form for the fishermen to use should they require more information. This website essentially represents their software application.

To start the build-out of the application, the first step is to formalize a team. The following diagram illustrates what the team looks like:

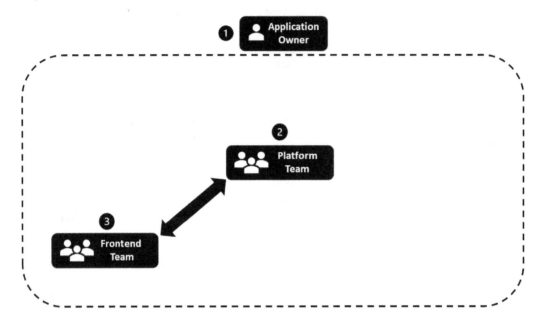

Figure 10.3 – The application team

As you can see from *Figure 10.3*, the initial application team is comprised of the following key resources:

1. Application owner
2. Site reliability/platform engineering team
3. Frontend application software engineering team

Let's examine these specific roles in more detail, starting with the application owner.

Examining the role of the application owner

The application owner's primary responsibility is to ensure that the website is strategically aligned with the goals of the business. Alongside this, the owner must ensure that the website is functional, usable, dependable, and operates in a cost-effective manner.

While the application owner may not be responsible for directly managing the platform or development engineers, they are responsible for directing and coordinating the various efforts performed by these teams. The application owner primarily owns the planning and design phase within the context of the MLSDLC. Some of their tasks may also include the following:

- **Documentation**: The application owner creates and manages application documentation to provide the engineering teams with the correct requirements and overall expectations.

- **Relationship management and strategic alignment**: The application owner coordinates feedback from users and other stakeholders to determine the product strategy, feature enhancements, and consistent alignment with business goals.

- **Analysis and reporting**: The application owner generates the necessary reports to communicate with the various stakeholders.

These are just a few of the tasks that may be performed by the application owner. The next role we will examine is that of the platform engineers.

Examining the role of the platform engineers

In previous chapters, we have often seen how the application development teams or **development-operations** (**DevOps**) teams have interacted with ML partitions to deliver the ML model into production. In the context of an MLSDLC, these teams are also referred to as the *platform* or *site reliability team*. Here, the platform team is responsible for designing the overall platform architecture (in conjunction with the application owner), building out the infrastructure, and maintaining the platform.

To demonstrate this, we are going to start the MLSDLC build-out using the AWS Cloud9 **integrated development environment** (**IDE**) to construct the ACME website as an AWS **Cloud Development Kit** (**CDK**) project, using the following steps:

1. Log in to the same AWS account you've been using, and open the AWS Cloud9 console (https://console.aws.amazon.com/cloud9).

2. In the **Your environments** section, click the **Open IDE** button for the MLOps-IDE development environment.

3. Run the following command in the Cloud9 terminal window to confirm that we are running version 2.3.0 (build beaa5b2) of the AWS CDK. Update the environment by running the following command in the workspace terminal:

```
$ cdk --version
```

> **Note**
>
> If you are not running version 2.3.0 (build beaa5b2) of the AWS CDK, refer to *Chapter 4, Continuous Integration and Continuous Delivery (CI/CD) for Machine Learning* for instructions on how to install this version.

4. Run the following commands to initialize and bootstrap the CDK application:

```
$ cd ~/environment
$ mkdir acme-web-application && cd acme-web-application
$ cdk init --language python
$ git add -A
$ git commit -m "Started CDK Project"
$ git branch main
$ git checkout main
$ source .venv/bin/activate
```

5. Next, we will install the necessary development libraries by running the following command:

```
$ python -m pip install -U pip pylint boto3
```

6. Since we will be making use of some experimental CDK construct libraries, using the left-hand navigation panel of the Cloud9 IDE, expand the `acme-web-application` folder and double-click on the `requirements.txt` file for editing, and then add the following alpha modules to the file:

```
aws-cdk.aws-apigatewayv2-alpha==2.3.0a0
aws-cdk.aws-apigatewayv2-integrations-alpha==2.3.0a0
```

7. Save and close the `requirements.txt` file.

8. Now, we install the required CDK modules by running the following command in the Cloud9 terminal window:

```
$ pip install -r requirements.txt
```

9. Now that we have the relevant libraries installed, we can start defining a skeleton CDK pipeline. Using the left-hand navigation panel, expand the `acme-web-application` folder and delete the `acme_web_application_stack.py` file.

10. Now, right-click on the `acme_web_application` folder and select the **New File** option to create a new file called `acme_pipeline_stack.py`. Double-click on the `acme_pipeline_stack.py` file for editing.

> **Note**
>
> You can reference the companion GitHub repository (https://github.
> com/PacktPublishing/Automated-Machine-Learning-on-
> AWS/blob/main/Chapter10/Files/cdk/acme_pipeline_
> stack.py) for a completed version of the acme_pipeline_stack.py
> file.

11. Add the following code to import the necessary libraries:

```
import aws_cdk as cdk
import aws_cdk.aws_codecommit as codecommit
import aws_cdk.aws_s3 as s3
import aws_cdk.pipelines as pipelines
import aws_cdk.aws_ssm as ssm
from constructs import Construct
```

12. Now, use the following code to initialize the PipelineStack class as a cdk.
Stack construct:

```
class PipelineStack(cdk.Stack):


    def __init__(self, scope: Construct, id: str, *,
model_name: str=None, group_name: str=None, repo_name:
str=None, feature_group: str=None, threshold: float=None,
cdk_version: str=None, **kwargs) -> None:
        super().__init__(scope, id, **kwargs)
```

13. The first resource we create is a CodeCommit source code repository, for all
the MLSDLC source code. Using the following code, we also create an output
for the **Uniform Resource Locator** (**URL**) so that other teams can easily clone
this repository:

```
        self.code_repo = codecommit.Repository(
            self,
            "Source-Repository",
            repository_name=repo_name,
            description="ACME Web Application Source Code
Repository"
        )
        cdk.CfnOutput(
            self,
```

```
            "Clone-URL",
            description="CodeCommit Clone URL",
            value=self.code_repo.repository_clone_url_
    http
        )
```

14. The next resource we create is a **Simple Storage Service (S3)** bucket to house all of the relevant ML and pipeline data. Here's the code to accomplish this:

```
        self.data_bucket = s3.Bucket(
            self,
            "Data-Bucket",
            bucket_name=f"data-{cdk.Aws.REGION}-{cdk.Aws.
    ACCOUNT_ID}",
            block_public_access=s3.BlockPublicAccess.
    BLOCK_ALL,
            auto_delete_objects=True,
            removal_policy=cdk.RemovalPolicy.DESTROY,
            versioned=True
        )
```

15. Next, we save the S3 bucket name, as well as the SageMaker Feature Store `FeatureGroup` name as Systems Manager parameters. These will be used by other teams to reference assets outside of the pipeline. Here's how we do this:

```
        ssm.StringParameter(
            self,
            "Data-Bucket-Parameter",
            parameter_name="DataBucket",
            description="SSM Parameter for the S3 Data
    Bucket Name",
            string_value=self.data_bucket.bucket_name
        )

        ssm.StringParameter(
            self,
            "Feature-Group-Parameter",
            parameter_name="FeatureGroup",
            description="SSM Parameter for the SageMaker
```

```
Feature Store group",
            string_value=feature_group
    )
```

16. Now, we create a `source_artifact` variable to essentially tell the CI/CD pipeline where to find the source code for the various artifacts and resources. The code is illustrated in the following snippet:

```
source_artifact = pipelines.CodePipelineSource.
code_commit(
        repository=self.code_repo,
        branch="main"
    )
```

17. Finally, we create a skeleton pipeline, using the following code:

```
pipeline = pipelines.CodePipeline(
        self,
        "Application-Pipeline",
        pipeline_name="ACME-WebApp-Pipeline",
        self_mutation=False,
        cli_version=cdk_version,
        synth=pipelines.ShellStep(
            "Synth",
            input=source_artifact,
            commands=[
                "printenv",
                f"npm install -g aws-cdk@{cdk_
version}",
                "python -m pip install --upgrade
pip",
                "pip install -r requirements.txt",
                "cdk synth"
            ]
        )
    )
```

18. Save and close the `acme_pipeline_stack.py` file.

19. Using the left-hand navigation panel, open the `app.py` file for editing.

20. Once the file is open, delete any existing template code and add the following code. In the app.py file, we are initializing the global parameters used by the CDK application, such as the name of the model, the name of the CodeCommit repository, and a placeholder for the name of the Feature Store. Additionally, in this code, we also instantiate the acme_pipeline_stack.py file as the PipelineStack() construct, within the CDK application:

```python
#!/usr/bin/env python3
import os
import aws_cdk as cdk
from acme_web_application.acme_pipeline_stack import PipelineStack


MODEL = "abalone"
MODEL_GROUP = f"{MODEL.capitalize()}PackageGroup"
FEATURE_GROUP = "PLACEHOLDER"
CODECOMMIT_REPOSITORY = "acme-web-application"
CDK_VERSION = "2.3.0"
QUALITY_THRESHOLD = 3.1


app = cdk.App()


PipelineStack(
    app,
    CODECOMMIT_REPOSITORY,
    env=cdk.Environment(account=os.getenv("CDK_DEFAULT_ACCOUNT"), region=os.getenv("CDK_DEFAULT_REGION")),
    model_name=MODEL,
    repo_name=CODECOMMIT_REPOSITORY,
    group_name=MODEL_GROUP,
    feature_group=FEATURE_GROUP,
    cdk_version=CDK_VERSION,
    threshold=QUALITY_THRESHOLD,
)


app.synth()
```

21. Save and close the `app.py` file.

22. Now, we deploy the skeleton pipeline. Run the following commands in the IDE terminal window, to bootstrap the application:

```
$ export CDK_NEW_BOOTSTRAP=1
$ npx cdk bootstrap aws://${CDK_DEFAULT_ACCOUNT}/${CDK_
DEFAULT_REGION} \
    --cloudformation-execution-policies
arn:aws:iam::aws:policy/AdministratorAccess
```

23. Deploy the CDK application, using the following command:

```
$ cdk deploy
```

24. Once the application has been deployed, and therefore the CodeCommit repository has been created, we can now check in the code so that other teams can access these resources. Run the following commands to initialize and update the repository:

```
$ CLONE_URL=$(aws cloudformation describe-stacks
--stack-name acme-web-application --query "Stacks[0].
Outputs[?OutputKey=='CloneURL'].OutputValue" --output
text)
$ git remote add origin $CLONE_URL
$ git add -A
$ git commit -m "Initial commit"
$ git push --set-upstream origin main
```

Now that the CDK application has been deployed, you will have noticed that we have bootstrapped it differently from the CDK examples in previous chapters. This is because we are using **CDK Pipelines**. CDK Pipelines essentially allows us to create a self-mutating CI/CD pipeline that deploys **CDK stacks** as pipeline **stages**. The pipeline is *self-mutating* in that it automatically builds the various pipeline assets, and dynamically adjusts its workflow when CDK constructs are added, updated, or deleted.

> **Note**
>
> For more information on CDK Pipelines, you can reference the launch blog (`https://aws.amazon.com/blogs/developer/cdk-pipelines-continuous-delivery-for-aws-cdk-applications/`).

As we will see, this concept fits nicely into the MLSDLC process, in that it essentially allows each cross-functional team to create, own, and dynamically provision the relevant AWS assets that pertain to their contribution to the overall MLSDLC process as a CDK construct.

> **Note**
>
> Since multiple cross-functional teams will be adding their contribution to the code source in this and the next chapter, in *Step 17*, we have set the `self_mutation` parameter to `False`. This will prevent the pipeline from self-mutating until all the relevant code contributions have been made.

This completes the *current* code contribution from the platform team. As *Figure 10.3* shows, the next role we will examine is that of the frontend development team.

Examining the role of the frontend developers

Since the ACME application involved deploying a website, the MLSDLC process requires a team of web developers. Per *Figure 10.3*, we see that the web developers interact with the application owner to determine the required look and feel of the website—essentially, the website design specifications.

This team also interacts with the platform team to create the necessary interfaces between the frontend and backend services.

> **Note**
>
> Due to the complexities within some of the MLSDLC assets, and to ensure consistency within this example, the majority of the code has already been created for you in the GitHub repository (`https://github.com/PacktPublishing/Automated-Machine-Learning-on-AWS/tree/main/Chapter10`). Therefore, some of the tasks within this chapter involve simply copying the code from the companion repository. You will recall that we have already cloned supporting code into the Cloud9 workspace in *Chapter 4, Continuous Integration and Continuous Delivery (CI/CD) for Machine Learning*.

Let's review how this team contributes to the MLSDLC process by creating the application's web assets. Follow these next steps:

1. Continuing in the Cloud9 environment, run the following commands within the IDE terminal window to move the website artifacts to the repository we created in the previous section, by running the following command:

```
$ cd ~/environment
$ cp -R src/Chapter10/www acme-web-application/
```

2. Commit these new website files to the MLSDLC source repository by running the following commands:

```
$ cd ~/environment/acme-web-application
$ git add -A
$ git commit -m "Added website files"
$ git push
```

By committing the various **HyperText Markup Language** (HTML) files, **Cascading Style Sheets** (CSS) files, and website **images**, this essentially completes the frontend developer team's contribution to the MLSDLC.

Within these various website assets, the most important HTML file to take note of is the index.html file. At the end of the code within the index.html file, you will see two JavaScript functions—namely, the submitContactForm() and submitPredictForm() functions. By developing these functions in conjunction with the platform team, the frontend team can ensure that the appropriate backend resources are created to support the data submitted within these forms—one for the contact processing and one for *Age Calculator* predictions.

Once the platform team creates these backend resources, it's at that point that we effectively have a completed web application. However, we still need to create ML components for the application. Let's explore these components further from the perspective of the ML practitioner and data engineering teams.

Examining ML and data engineering roles

In previous chapters, we have used the term **ML practitioner** as a blanket term for any person responsible for automating the ML process. Within the context of the MLSDLC process, we typically see this role split into two distinct functions, namely the following:

- **Data scientist**: The data scientist is primarily responsible for building, training, and tuning an ML model that meets the business requirements of the use case.

- **ML engineer**: Among numerous responsibilities, the ML engineer is primarily responsible for designing the overall ML system to support the model, managing the appropriate datasets for model training, and ensuring the final ML application addresses the business requirements for the use case.

However, for the sake of the ACME application example, we will group these two functions under the banner of the ML team, with the following diagram highlighting how this team fits into the MLSDLC process:

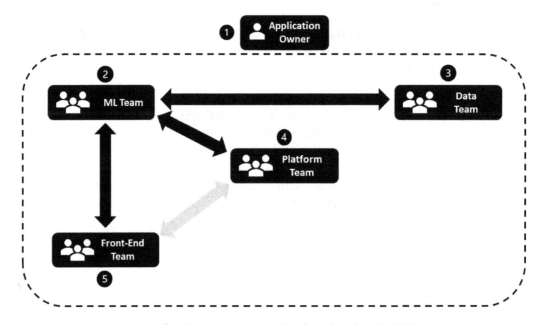

Figure 10.4 – The ML practitioner team's role within the MLSDLC process

From *Figure 10.4*, you can see that the application owner works with the ML team to assess whether ML can be applied to the business case. You will recall that we reviewed this process in more detail when making a case for ML in *Chapter 1*, *Getting Started with Automated Machine Learning on AWS*.

So, once it has been determined that ML is a fit for the business case, the next step is to determine whether we have *supporting data* for the ML model. It's at this point that the ML engineers and data engineers coordinate on the data source, access requirements, type of data, and how the data needs to be re-engineered for the ML model. Since the ACME use case uses data from the **University of California Irvine (UCI)** *Machine Learning Repository*, we are going to forego this step of interfacing with the data engineers.

However, within the context of the MLSDLC, we are going to introduce a technique, typically performed by ML engineers, to further streamline training data processing and **feature engineering (FE)** tasks—using a **Feature Store**.

Creating a SageMaker Feature Store

At *re:Invent 2020*, AWS launched a SageMaker capability called the **SageMaker Feature Store**. This allows teams to create, store and reuse preprocessed and engineered features, essentially eliminating the need to constantly execute data preprocessing jobs and FE tasks every time a model needs to be trained.

> **Note**
>
> AWS provides several example notebooks for creating a Feature Store, within the Amazon SageMaker example GitHub repository (`https://github.com/aws/amazon-sagemaker-examples/tree/master/sagemaker-featurestore`). We will be reusing code from these examples, licensed under the **Apache 2.0** license, and adapting them to our use case.

The following steps will take us through creating a store for the abalone dataset features:

1. Within your AWS account, open the **Amazon SageMaker** management console, and click the **SageMaker Domain** link in the left-hand navigation panel.

2. Click the **Launch app** dropdown, and select the **Studio** link to launch the Studio **user interface (UI)**.

3. Since the companion GitHub repository has already been cloned into the Studio UI, using the **File Browser** panel, double-click on the `src` folder, and then double-click on the `Chapter10` folder, then the `Notebooks` folder.

4. Now, double-click on the `SageMaker Feature Store Example.ipynb` notebook to launch it.

5. Once the notebook is open, wait for the notebook kernel to start.

6. Select **Kernel** from the menu bar, and select the **Restart Kernel and Run All Cells…** option.

7. When prompted on the **Restart Kernel?** dialog box, click the **Restart** button.

> **Note**
>
> The notebooks should take around 10 minutes to run. Make sure to take note of the name of the feature group, as we will be referencing this in the next chapter.

Let's review what we've accomplished from executing this notebook, as follows:

- In the **Setup** section of the notebook, we import the necessary Python libraries, with `sagemaker.feature_store.feature_group` being the most important. We've also declared some helper functions to track the status of the feature stores' creation using the `check_feature_group_status()` function, and tracked the ingestion of the feature data into the store, using the `check_data_availabiltiy()` function. You will also see that we reference the S3 data bucket (`data_bucket`) that was created by the platform team in the *Examining the role of the platform engineers* section, by pulling the bucket name parameter from the **Systems Manager Parameter Store (SSM)**.

- In the **Data Preparation** section, we download the abalone dataset from the UCI repository and, using the pandas `get_dummies()` method, we engineer the sex features as numerical values. We then store these new features as a DataFrame called `processed_data`.

- Finally, in the **SageMaker Feature Store** section, we create a feature group, which is essentially a table within the Feature Store. We also create a `time_stamp` variable to bind an ingestion timestamp to our data as a feature column. This allows us to differentiate between individual features, based on the time they were added to the group. We then define a schema for the feature group, create it, and ingest the `processed_data` DataFrame into the table.

After running the notebook, we now have a Feature Store with all the relevant abalone dataset features, thus eliminating the need to constantly recreate these features every time we train our model.

So, now that we have the dataset ready, we can move on to creating ML artifacts.

Creating ML artifacts

From *Figure 10.4*, we can see that after coordinating with the data team, the ML team works with the platform team to convey its requirements and provide the ML-specific code contributions to the web application, the first of which is the model artifact.

Creating a model artifact

You will recall from *Chapter 5, Continuous Deployment of a Production ML Model,* that we packaged the algorithm code, as well as various routines to process the data and train and evaluate the model into a container image. This allowed us to compile an all-inclusive model artifact for the various stages of the CI/CD pipeline, using SageMaker's **Bring Your Own Container** (**BYOC**) capabilities.

Within the context of the MLSDLC example, after the data scientists have framed the correct ML solution, they can build, train, tune, and evaluate a production-grade model for the solution, essentially reproducing the model artifacts using the same notebook example that we used in *Chapter 4, Continuous Integration and Continuous Delivery (CI/CD) for Machine Learning.* The data scientists can then package these components into a container artifact for the ACME web application.

Let's emulate this assignment with the following steps:

1. Open the `ACME Model Artifacts Example.ipynb` file, using the SageMaker Studio IDE, in the `src/Chapter10/Notebooks` folder of the cloned companion GitHub repository.

2. Once the **Python 3 (Data Science)** kernel has started, select **Kernel** from the menu bar, and select the **Restart Kernel and Run All Cells...** option.

3. When prompted on the **Restart Kernel?** dialog box, click the **Restart** button.

4. After the notebook has been run, you should see a `model` folder in the left-hand navigation panel. This folder contains the relevant model artifacts for the container image.

5. Now, open a terminal by clicking **File** from the menu bar, selecting the **New** option, and then clicking on the **Terminal** option.

6. Run the following commands within the terminal tab to add the model artifacts to the web application source code repository:

```
$ CLONE_URL=$(aws cloudformation describe-stacks
--stack-name acme-web-application --query "Stacks[0].
Outputs[?OutputKey=='CloneURL'].OutputValue" --output
text)
$ git clone $CLONE_URL
```

```
$ mv ~/src/Chapter10/Notebooks/model acme-web-
application/
$ cd acme-web-application/
$ git add -A
$ git commit -m "Initial commit of model artifacts"
$ git push
```

As the ML team, we have now created the relevant model artifacts and contributed these to the web application repository. However, before proceeding to the next step, let's review what happened when we ran the notebook.

If you examine the notebook, you will see that we've followed a similar procedure to the example in *Chapter 5, Continuous Deployment of a Production ML Model*, whereby we use the %%writefile magic to create a model.py file. This file loads the necessary TensorFlow libraries, sets global variables for the SageMaker container environment, and defines a model training routine in the form of the train() function. This function defines training and validation data and a **multilayer perceptron** (**MLP**) model, executes the training fit() method on the compiled model, and then saves the optimized model.

In the **Create the Application** section of the notebook, we create an app.py file, which serves as the entry point to the container image for either the model training task or the model inference task, depending on how SageMaker consumes the image. In this section, we also initialize the web serving files, nginx.conf and wsgi.py, so that SageMaker can host and serve the model for inferencing.

The last section of the notebook creates a Dockerfile. This file contains the build instructions to create a container image. Unlike the previous example, we aren't pulling a **deep learning** (**DL**) container image. Instead, we are manually building a container image.

> **Note**
>
> The primary reason for manually building a TensorFlow container image, as opposed to pulling the DL container image, is to ensure that the code example works across any AWS regions that support CDK Pipelines and SageMaker. While the CDK Pipelines module supports the ability to supply docker_credentials within the aws_cdk.pipelines.CodePipeline() class, we would need to hardcode credentials to the DL container **Elastic Container Registry** (**ECR**) repositories within the example code. So, to ensure the example code works uniformly, we will manually build a container based on the DL container source (https://github.com/aws/deep-learning-containers/blob/master/tensorflow/training/docker/2.6/py3/Dockerfile.cpu), provided under the Apache 2.0 license.

Within the `Dockerfile`, you will also see that we install an additional Python library called `awswrangler` (`https://github.com/awslabs/aws-data-wrangler`). **AWS Data Wrangler** is an AWS-developed and open sourced library that allows easy integration with various AWS services, such as **Amazon Athena**, **AWS Glue**, and **Amazon Redshift**. Since the training data is housed within the SageMaker Feature Store, we will use this library to select the feature data and store this as a DataFrame for model training.

Developing the model artifacts doesn't complete the ML team's contribution to the MLSDLC example. If you recall from previous chapters, the ML team also needs to contribute various additional artifacts to automate the model building and evaluation process. Let's explore these additional artifacts in the next section.

Building automated ML workflow artifacts

In previous chapters, we've reviewed multiple techniques to automate the model training and evaluation process. For example, in *Chapter 6*, *Automating the Machine Learning Process using AWS Step Functions*, and *Chapter 8*, *Automating the Machine Learning Process using Apache Airflow*, you were introduced to some of the AWS capabilities that build a workflow to orchestrate getting an ML model into production. We also review the importance of having a cross-functional team co-develop these workflow artifacts, and not having the platform team own the entirety of these tasks.

So, within the context of the MLSDLC example, the ML team further contributes to the ACME web application by providing a codified workflow that gets executed as part of the CDK pipeline. To this end, let's walk through the process of building these artifacts, from the perspective of the ML engineers. Here are the steps we need to follow:

1. As the ML engineer, we need to update the cloned repository with the latest updates to the model artifacts. Using the Cloud9 IDE workspace terminal window, run the following commands:

    ```
    $ cd ~/environment/acme-web-application/
    $ git pull
    ```

2. In the left-hand navigation panel, create a folder called `stacks`, to hold the CDK constructs. You can do this by right-clicking on the `acme_web_application` folder and selecting **New Folder**. Then, name the folder `stacks`.

3. Copy the pre-built stack construct from the companion GitHub repository to this folder, by running the following command:

```
$ cp ~/environment/src/Chapter10/Files/cdk/ml_workflow_
stack.py acme_web_application/stacks/
```

4. Using the left-hand navigation panel, double-click on the `ml_workflow_stack.py` file for review.

In previous chapters, we've reviewed different ways to automate the ML workflow—namely AWS Step Functions and MWAA. At *re:Invent 2020*, AWS launched a native SageMaker module to orchestrate the ML process, called SageMaker Pipelines. As we review the `ml_workflow_stack.py` file, you will notice that we automate the ML process using AWS Step Functions instead of using SageMaker Pipelines, for two reasons. Firstly, you should already be familiar with using AWS Step Functions, from the Data Science SDK example in *Chapter 6, Automating the Machine Learning Process using AWS Step Functions*.

Secondly, while the CDK supports executing a SageMaker pipeline using the `CfnPipeline` construct (https://docs.aws.amazon.com/cdk/api/latest/python/aws_cdk.aws_sagemaker/CfnPipeline.html), this construct requires the pipeline to be separately codified and unit tested as an artifact, outside of the CDK project. In the next chapter, we will see that by integrating AWS Step Functions into the CDK project, the process of codifying, unit testing, and—eventually—system testing the ML workflow can be further automated, as part of the self-mutating CDK pipeline.

Now that the `ml_workflow_stack.py` file is open, let's review the most important AWS resources created by the stack construct.

Registering the data bucket

Outside of loading the necessary CDK Python libraries and instantiating the `MLWorkflowStack()` class, the first variable we declare is the `data_bucket` variable. Here, we reference the existing S3 bucket as a CDK object, thus allowing us to add the various permissions required by the other stack resources to add and access the objects within the bucket. For example, we use the `BucketDeployment()` construct next, to upload the Python script artifacts to the `data_bucket` variable so that these objects can be used with the workflow.

Creating placeholder parameters

Next, we declare two SSM parameters (`package_parameter` and `baseline_parameter`) as placeholders, to store the name of the trained model and the S3 bucket location of the SageMaker Model Monitor baseline data.

Creating a modeling container image

Since we now have the model artifacts already created, we can define the `model_image` parameter as an ECR `DockerImageAsset()` asset, pointing to the model artifacts folder. As you will see, by declaring this asset, the self-mutating CDK pipeline will dynamically create a CodeBuild job to build a container image, without us having to declare a separate CodePipeline build stage.

Creating a model registry

Next, we create an AWS Lambda function variable called `registry_creator`. This Lambda function creates a SageMaker model registry (`https://docs.aws.amazon.com/sagemaker/latest/dg/model-registry.html`) to store the various model versions that are automatically trained when the workflow gets executed. Once the `registry_creator` Lambda function has been declared, we invoke it as a custom resource using the `CustomResource()` construct.

> **Note**
>
> While the CDK provides a SageMaker construct called `CfnModelPackageGroup` (`https://docs.aws.amazon.com/cdk/api/latest/python/aws_cdk.aws_sagemaker/CfnModelPackageGroup.html`) to register the trained model package, we use a Lambda function here to essentially perform the same task. As you will see when we define a Lambda artifact later in this chapter, using a Lambda function will allow us to delete existing model packages before deleting the model registry—something the `CfnModelPackageGroup` construct doesn't do.

Creating an ML experiment

Before we can define a workflow as a Step Functions state machine, we need to define artifacts that will be used within the workflow. The first artifact is the `experiment_creator` **Lambda function**. This function initializes the experiment variables, tagged with the pipeline `executionId` for version tracking so that each execution of the workflow can be traced. This allows the ML team to track the lineage of a production model, from the data used for training to how the model was trained and how the model was evaluated, as a SageMaker experiment. This information is useful for auditing purposes and model explainability and provides additional context for production model monitoring.

Evaluating the model

The next artifact definition is the `evaluate_results` Lambda function. This function reads the model evaluation results from the *current* model being trained within the workflow with the evaluation results from any *previously* trained model, to determine whether or not the *current* model's performance is an improvement. This way, we can guarantee that the production model is always the best-performing model and doesn't get overridden by an inferior model. Should the model's performance improve, we use the `register_model` Lambda function to update the model registry with the latest, best model. This is the model that will eventually be deployed into production.

Creating SageMaker jobs

Now that the Lambda artifacts of the workflow have been defined, we can build various SageMaker **application programming interface (API)** calls. There are three specific API calls to SageMaker, as follows:

- `CreateProcessingJob` (https://docs.aws.amazon.com/sagemaker/latest/APIReference/API_CreateProcessingJob.html) for data processing, using the `processing_definition` state JSON

- `CreateTrainingJob` (https://docs.aws.amazon.com/sagemaker/latest/APIReference/API_CreateTrainingJob.html) for model training, using the `training_definition` state JSON

- `CreateProcessingJob`, for model evaluation, using the `evaluation_definition` state JSON

To define the API specification for each of these jobs, we've created three separate definition parameters in the workflow construct—namely, the following:

- `processing_definition`
- `training_definition`
- `evaluation_definition`

> **Note**
>
> Since we've already created a Feature Store to hold the engineered data features, you may be wondering why we've included a *data processing definition* in the workflow. While the Feature Store does contain the training data, we still need to split the data into training, validation, and test datasets. Therefore, by using the `processing_definition`, we are offloading the task of retrieving the engineered data from the Feature Store, splitting the data into training, validation, and test datasets, and storing them in S3.

Defining a state machine

Now that we have defined the various workflow artifacts and SageMaker job definitions, we can declare the different tasks and states of the Step Functions state machine. You will recall from the example in *Chapter 4, Continuous Integration and Continuous Delivery (CI/CD) for Machine Learning* that we started the build-out of the flow by looking at the final objective state for the workflow, and then working backward to develop the steps toward accomplishing that objective. So, if we follow what we've already learned, the first state we create is `failure_state`, using the `Fail()` class of the `aws_stepfunctions` construct.

Next, we define steps that lead us to the final workflow objective, having a production-grade model, starting with `create_experiment_step`. As you can see, this variable is an `aws_stepfunctions_task` function called `LambdaInvoke()`, whereby we call the `experiment_creator` function to initialize the experiment parameters for workflow tracking.

The subsequent step is the data `processing_step` variable, through which we register `processing_definition` as a Step Functions `CustomState()`.

> **Note**
>
> You have probably noticed throughout the various examples in this book that the CDK is constantly being updated, and while there is a CDK construct for the SageMaker training job (`https://docs.aws.amazon.com/cdk/api/latest/python/aws_cdk.aws_stepfunctions_tasks/SageMakerCreateTrainingJob.html`), at the time of writing, there is currently no construct for SageMaker processing jobs. Therefore, define these workflow steps as a `CustomState()` function.

After `processing_step` comes the `training_step` variable, which uses `training_definition` to execute the SageMaker training job as a `CustomState()`.

> **Note**
>
> Even though the CDK has a `SageMakerCreateTrainingJob()` class, at the time of writing, this class does not support adding a SageMaker experiments configuration. Therefore, to add the lineage tracking capability to the training job, we've declared `training_step` as a `CustomState()`.

Once the model has been trained, we can evaluate its performance, using a SageMaker processing job. This step of the workflow is defined by the `evaluation_step` variable and uses the `evaluation_definition` variable to provide the necessary API configuration to SageMaker. The subsequent `results_step` variable then uses these evaluation metrics, by invoking the `evaluate_results` Lambda to determine whether or not the trained model is ready for production deployment.

After applying various workflow logic steps to direct the overall flow, we create a workflow as a Step Function state machine, using the `workflow_definition` variable. The following diagram depicts what the final Step Functions state machine will look like:

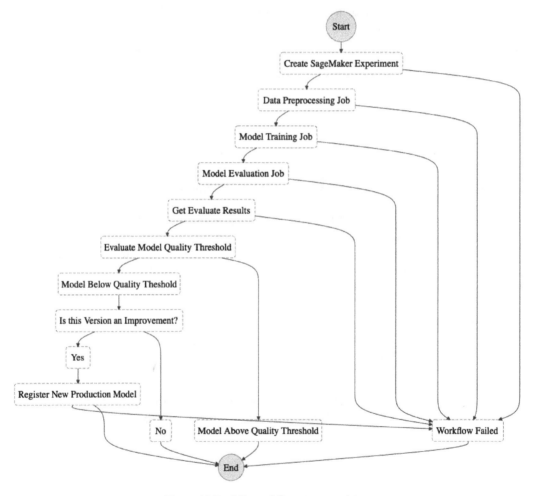

Figure 10.5 – ML workflow state machine

Finalizing the workflow artifacts

Once the workflow construct has been defined, the final part of the ML team's contribution to the web application is to supply the various artifacts referenced within the construct—namely, the artifacts in the `lambda` and `scripts` folders. The following steps will take you through creating these supporting artifacts:

1. Continuing within the Cloud9 IDE, run the following commands in the terminal window, to add the code for the `registry_creator`, `experiment_creator`, `evaluate_results`, and `register_model` Lambda functions:

```
$ cd ~/environment/acme-web-application
$ mkdir lambda
$ cp -R ~/environment/src/Chapter10/Files/lambda/
{createExperiment,evaluateResults,registerModel,
registryCreator} lambda/
```

2. Now, execute the following commands to copy the required scripts for the `processing_step` and `evaluation_step` variables:

```
$ cd ~/environment/acme-web-application
$ cp -R ~/environment/src/Chapter10/Files/scripts .
```

3. Commit these changes to the web application source repository, as follows:

```
$ git add -A
$ git commit -m "Initial commit of ML Workflow artifacts"
$ git push
```

You can review each of the `index.py` files within the individual function's folder, to assess exactly what the function does within the ML workflow. However, you should pay particular attention to the `preprocessing.py` file in the `scripts` folder, to see how AWS Data Wrangler reads the feature data from the Feature Store. For example, if you refer to the following code snippet, you can see that AWS Data Wrangler performs a **Structured Query Language (SQL)** query against the raw Feature Store data using the Amazon Athena (https://aws.amazon.com/athena/) service:

```
...
if __name__ == "__main__":

    ...

    query_string = f'SELECT {",".join(columns)} FROM "{table}"
WHERE is_deleted=false;'
    featurestore_df = wr.athena.read_sql_query(query_string,
```

```
database=database, ctas_approach=False)
   ...
   X = shuffle(featurestore_df).to_numpy()
   ...
   training, validation, testing = np.split(X,
[int(.8*len(X)), int(.95*len(X))])
   ...
```

Since the Feature Store is essentially a metastore for the raw feature data, which is stored in Parquet files in S3, Athena can be used to perform interactive SQL queries directly against the data. As you can see from the highlighted code snippet, we use AWS Data Wrangler to select the relevant feature columns using the `athena.read_sql_query()` method, and store the results as a DataFrame. The code continues to shuffle the data, removing any ordered indexing from the query, and splits the data into specific training, validation, and test datasets.

So, after running these previous steps as the ML team, we have officially contributed the required ML artifacts to the ACME web application, and therefore we can sign off on the interactions with the platform team. However, there is still one more group that the ML team needs to interact with—the frontend developers.

Adding ML to the frontend application

In *Figure 10.4*, we can see that the final interaction that the ML team has within the context of the MLSDLC is with the frontend team. During this engagement, these two teams determine which web UI changes need to be made in order for the web application user to make inferences against the production ML model—in essence, how users will inevitably use the *Age Calculator*.

This requires the web developers to create an HTML form, whereby fishermen can enter the physical measurements of the abalone into the web UI and have the production-grade ML model predict the age.

> **Note**
> The HTML form code and supporting JavaScript function have already been provided for you. You can review this code by referencing the `predictionModel` HTML code, and `submitPredictForm()` JavaScript code, in the `index.html` file.

After completing the necessary code updates, verifying that they meet the functional requirements outlined by the application owner, and committing these into the application source code repository, the ML team can sign off on its contribution to the ACME web application.

If our example application were based solely on an SDLC process, we technically have all the artifacts necessary to update the skeleton CDK pipeline and deploy the web application using the CI/CD process. However, since we are creating an ML-based SDLC, there is one final component that we need to incorporate into the overall automation process. Let's explore what this is, in the next section.

Creating continuous training artifacts

In *Chapter 8, Automating the Machine Learning Process using Apache Airflow*, you were introduced to a fundamental requirement for any ML automation initiative— that is, the ability to automatically re-train an ML model with new data. In the same chapter, we also demonstrated this requirement by showing how data engineers can use the MWAA service to orchestrate a data-centric workflow to train the *Age Calculator* model on updated survey data.

Even though we are focusing on the MLSDLC process in this chapter, we still have the business requirement to incorporate new survey data into the ACME web application example. *So, how do we address this business requirement within the context of the current CI/CD pipeline?*

The answer to this question is relatively simple. Since we are using a CI/CD pipeline to automate the delivery of our web application (along with a production-grade ML model), we can simply apply what we've already learned, and tack on the requirement to re-train the ML model after it has been deployed into production. This is essentially the premise behind **continuous training** (**CT**), whereby we add the ability to restart the CI/CD process (once we have new data) to create an automated **CI/CD/CT** methodology.

Since the data engineering team was responsible for delivering the data-centric workflow example in *Chapter 8, Automating the Machine Learning Process using Apache Airflow*, we can further extend their role within the MLSDLC example to provide CT artifacts. The following diagram illustrates the overall role that the data team plays within the MLSDLC process:

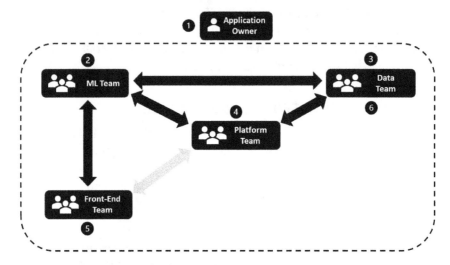

Figure 10.6 – The data team's role with the MLSDLC process

Now that we have started to get a picture of the overall role of the data team, let's dive into the artifacts they contribute to the process.

Building data workflow artifacts

Following the same procedures as the ML team, the data team contributes to the ACME web application by providing the necessary MWAA infrastructure components that get executed as part of the CDK pipeline. Let's walk through the process of building these artifacts, from the perspective of the data engineers, as follows:

1. As a data engineer, open the Cloud9 IDE workspace, and using the terminal window, copy the pre-built stack construct from the book's GitHub repository, by running the following command:

```
$ cd ~/environment/acme-web-application/
$ cp ~/environment/src/Chapter10/Files/cdk/data_workflow_
stack.py acme_web_application/stacks/
```

2. Using the left-hand navigation panel, double-click on `data_workflow_stack.py` for review.

Once the `data_workflow_stack.py` file is open, we can review the important infrastructure resources created by the stack constructs.

As you can see, after importing the required CDK libraries, we instantiate the `DataWorkflowStack()` class as a `cdk.Stack` construct. The first thing we do is register the `data_bucket`, `data_bucket_param`, and feature group SSM parameter (`group_name_param`) variables. We do this so that we can assign the relevant access permissions to `airflow_role`.

After defining `airflow_role` and the appropriate `airflow_policy_document` variable (https://docs.aws.amazon.com/mwaa/latest/userguide/mwaa-create-role.html), we build out the **Virtual Private Cloud** (**VPC**) since MWAA requires a VPC, plus various networking components to support an environment.

> **Note**
>
> You will recall from *Chapter 8, Automating the Machine Learning Process using Apache Airflow,* that we created an MWAA VPC stack using the provided CloudFormation template. In this example, we are codifying the same network environment using the CDK.

Next, we instantiate the MWAA environment as the `airflow_environment` variable and create an S3 deployment construct to upload the Airflow DAG artifacts to S3.

Finally, we create a Lambda function called `releaseChange` to call the CodePipeline service and start a pipeline execution.

Now that the CT resources have been defined as a CDK construct, the next task for the data team to complete is to build the various artifacts that the construct references—namely, the `releaseChange` Lambda code and the Airflow DAG. The following steps will show you how to do this:

1. Using the terminal windows of the Cloud9 IDE, run the following command to copy the pre-built `releaseChange` Lambda code artifacts:

```
$ cd ~/environment/
$ cp -R ~/environment/src/Chapter10/Files/lambda/
releaseChange acme-web-application/lambda/
```

> **Note**
>
> You can review the `index.py` file, in the `acme-web-application/` `lambda/releaseChange` folder, to see how the Lambda function uses the `start_pipeline_execution()` API call to trigger a CodePipeline execution.

2. Using the left-hand navigation panel, of the Cloud9 workspace, right-click on the `acme-web-application` folder and select the **New Folder** option.

3. Create a folder called `airflow`.

4. Right-click on the newly created `airflow` folder and select the **New File** option.

5. Create a file called `requirements.txt` and double-click on it for editing.

6. Add the following code to the `requirements.txt` file:

```
sagemaker==2.49.1
s3fs<=0.4
boto3>=1.17.4
numpy
pandas
```

> **Note**
>
> The only reason we specifically reference version `2.49.1` of the SageMaker Python SDK is to ensure uniformity across all examples within the book.

7. Save and close the `requirements.txt` file.

8. Right-click on the `airflow` folder and select the **New Folder** option.

9. Create a folder called `dags`.

10. Right-click on the newly created `dags` folder and select the **New File** option.

11. Create a file called `continuous_training_pipeline.py` and double-click on the file for editing.

12. Add the following code to import the required Python libraries, in order to access the Feature Store:

```
import time
import json
import sagemaker
import boto3
import numpy as np
```

```
import pandas as pd
from time import sleep
from datetime import timedelta
from sagemaker.feature_store.feature_group import
FeatureGroup
```

13. Next, add the required Apache Airflow libraries to construct the DAG, as follows:

```
import airflow
from airflow import DAG
from airflow.operators.python_operator import
PythonOperator
from airflow.providers.amazon.aws.hooks.lambda_function
import AwsLambdaHook
from airflow.providers.amazon.aws.sensors.s3_prefix
import S3PrefixSensor
```

14. Now, add the following code to create global variables to reference the S3 `data_` `bucket`, the `releaseChange` Lambda Function, and the feature group name (`fg_name`) parameters from SSM:

```
sagemaker_session = sagemaker.Session()
region_name = sagemaker_session.boto_region_name
data_bucket = f"""{boto3.client("ssm", region_
name=region_name).get_parameter(Name="DataBucket")
["Parameter"]["Value"]}"""
data_prefix = "abalone_data"
lambda_function = f"""{boto3.
client("ssm", region_name=region_name).get_
parameter(Name="ReleaseChangeLambda")["Parameter"]
["Value"]}"""
fg_name = f"""{boto3.client("ssm", region_name=region_
name).get_parameter(Name="FeatureGroup")["Parameter"]
["Value"]}"""
```

15. Now, add the following code to initialize the Airflow DAG default configuration:

```
default_args = {
    "owner": "airflow",
    "depends_on_past": False,
    "start_date": airflow.utils.dates.days_ago(1),
```

```
        "retries": 0,
        "retry_delay": timedelta(minutes=2)
}
```

16. Next, we define code that gets executed within each step of the Airflow DAG, as Python functions. The first function (called `start_pipeline`) calls the `releaseChange` Lambda function to trigger an execution of the CI/CD/CT pipeline. Define this function with the following code:

```
def start_pipeline():
    hook = AwsLambdaHook(
        function_name=lambda_function,
        aws_conn_id="aws_default",
        invocation_type="RequestResponse",
        log_type="Tail",
        qualifier="$LATEST",
        config=None
    )
    request = hook.invoke_lambda(payload="null")
    response = json.loads(request["Payload"].read().
decode())
    print(f"Response: {response}")
```

17. The next function (called `update_feature_group`) takes the newly added abalone survey data, encodes the sex feature as numerical data, creates a `time_stamp` variable, and ingests this new data into the Feature Store. The code is illustrated in the following snippet:

```
def update_feature_group():
    fg = FeatureGroup(name=fg_name, sagemaker_
session=sagemaker_session)
    column_names = ["sex", "length", "diameter",
"height", "whole_weight", "shucked_weight", "viscera_
weight", "shell_weight", "rings"]
    abalone_data = pd.read_csv(f"s3://{data_bucket}/
{data_prefix}/abalone.new", names=column_names)
    data = abalone_data[["rings", "sex", "length",
"diameter", "height", "whole_weight", "shucked_weight",
"viscera_weight", "shell_weight"]]
    processed_data = pd.get_dummies(data)
```

```
    time_stamp = int(round(time.time()))

    processed_data["TimeStamp"] = pd.Series([time_stamp]
* len(processed_data), dtype="float64")

    fg.ingest(data_frame=processed_data, max_workers=5,
wait=True)

    sleep(300)
```

> **Note**
>
> The `update_feature_store` function essentially performs the same tasks as the code created by the ML engineers in the `SageMaker Feature Store Example.ipynb` notebook, except the previous code updates the existing feature group with the newer abalone survey data, as opposed to the original data that was downloaded from the UCI *Machine Learning Repository*.

18. Now that task execution functions have been defined, we can create a DAG workflow. The following code initializes a DAG called `acme-data-workflow` with the default arguments:

```
with DAG(
    dag_id=f"acme-data-workflow",
    default_args=default_args,
    schedule_interval="@daily",
    concurrency=1,
    max_active_runs=1,
) as dag:
```

19. The first step with the DAG uses the `S3PrefixSensor()` provider class to watch the S3 data bucket for any new data. The code is illustrated in the following snippet:

```
s3_trigger = S3PrefixSensor(
    task_id="s3_trigger",
    bucket_name=data_bucket,
    prefix=data_prefix,
    dag=dag
)
```

20. Once new survey data is uploaded to S3, Airflow executes the second step of the DAG by executing the `update_feature_group` function, using the Airflow `PythonOperator()` provider. The code is illustrated in the following snippet:

```
update_fg_task = PythonOperator(
    task_id="update_fg",
    python_callable=update_feature_group,
    dag=dag
)
```

21. The final step of the workflow is to trigger a release change of the CI/CD/CT pipeline by calling the `releaseChange` Lambda function to start a CodePipeline execution. The Airflow step accomplishes this task by calling the `start_pipeline` function, using the `PythonOperator()` provider. The code is illustrated in the following snippet:

```
trigger_release_task = PythonOperator(
    task_id="trigger_release_change",
    python_callable=start_pipeline,
    dag=dag
)
```

22. Now that the DAG steps have been defined, the last part of the code chains them together to finalize the DAG, as follows:

```
s3_trigger >> update_fg_task >> trigger_release_task
```

23. Save and close the `continuous_training_pipeline.py` file.

24. Run the following commands to commit the data team's contribution to the ACME web application repository:

```
$ cd ~/environment/acme-web-application
$ git add -A
$ git commit -m "Initial commit of CT artifacts"
$ git push
```

By completing the preceding steps, verifying that they meet the functional requirements as outlined by the application owner and committing these into the application source code repository, the data team can sign off on its contribution to the ACME web application.

At this point in the example, all the pertinent contributions from the cross-functional team have been developed, and the ACME web application is almost ready for deployment. However, there is one more team that needs to weigh in before the solution can be deployed. Let's explore this team's role in the next section.

Understanding the security lens

Securing the solution is a critically important task within the MLSDLC. While the majority of common MLSDLC implementations typically deal with security issues as and when they arise, it is a good practice to proactively assess the overall security posture of the final application before it's deployed into production.

So, instead of performing a full security audit on the ACME web application, this section will highlight some of the best practices that the security team should follow, by showing how they interact with other members of the cross-functional team. The following diagram shows the overall role that the security team plays within the MLSDLC process:

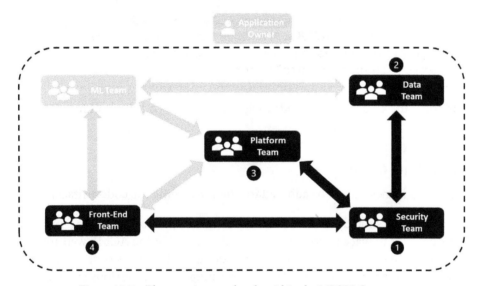

Figure 10.7 – The security team's role within the MLSDLC process

As *Figure 10.7* shows, the first thing the security team needs to do is review how data is used, by working with the data team.

Securing the data

The following guidelines should be followed when working with the data team to ensure that all data is secure:

- Any data, whether it's stored in a database or on a filesystem (on-premises or in the cloud), should be encrypted.

- Any data read from or written to these data stores should be encrypted.

- Any applications or people that access the data should be authorized to do so using the appropriate access controls. These access controls should include logging capabilities to ensure that authorized and unauthorized access can be traced and audited.

- No customer **personally identifiable information** (**PII**) with these data stores should be accessible by the data team or the ML team.

> **Note**
>
> For more information on these suggestions, and more, review the *Data Protection* section of the AWS *Best Practices for Security, Identity, & Compliance* web page (`https://aws.amazon.com/architecture/security-identity-compliance`).

The second thing the security team needs to do is review the code by working with the data team.

Securing the code

The following guidelines should be adhered to when reviewing the code artifacts:

- All private code should be in a secure source code repository, with the appropriate access controls in place to govern access to the code. For the ACME web application example, we use CodeCommit, which provides granular access controls to the repository and branch levels, while also governing access to various tasks that can be performed against the repository.

- There should be no application or user credentials, nor any passwords, in any of the code. These *secrets* should be stored in a separate store, such as AWS Secrets Manager (`https://aws.amazon.com/secrets-manager/`), where access can be controlled, logged, and audited.

In the case of the ACME web application example, securing the code is further compounded by the fact that the code creates AWS resources. Therefore, it is recommended that the security team also includes a member of the platform team, to create a **security-operations (SecOps) team**. This way, securing the code can extend to securing the provisioned AWS resources. For example, the CT artifacts create an MWAA infrastructure using a VPC. The SecOps team should review the VPC to ensure the following:

- All network traffic in and out of the VPC, as well as within the VPC, is logged and encrypted.

- All network ports are secured, using network **access control lists** (**ACLs**) and security groups.

- Any IAM roles created should, where possible, grant only the permissions required to perform the task required by the role.

Last, but not least, the security team must work with the website content developers to secure their respective artifacts.

Securing the website

As *Figure 10.7* highlights, the last group that the security team interacts with is the frontend team. Here are some suggestions for securing the website:

- All website content should only be accessible via a secure web server —in other words, the static content should be accessible via the appropriate URL and not directly accessible, say, from the S3 bucket.

- All traffic to and from the website should be encrypted with the **HyperText Transfer Protocol Secure** (**HTTPS**) protocol, using the appropriate **Secure Sockets Layer** (**SSL**) or **Transport Layer Security** (**TLS**) certificates.

- It is also recommended that the security team includes compliance resources to ensure that all content complies with regional or international accessibility standards.

- All public referenceable content or open source content must be documented and include the applicable license.

These preceding suggestions only cover a few of the focus areas for the security team as it pertains to the ACME web application, but once the security team concludes its review of the application artifacts and signs off, we are almost ready to deploy the ACME web application into production. All that's left to do is integrate every team's artifacts into the CDK pipeline. Once this task is completed by the platform team, the pipeline will be complete, and the application can be deployed into production. This will be the focus of the next chapter.

Summary

In this chapter, you were introduced to the concept of the MLSDLC, as a process that can be used to automate an E2E ML-based application. We also reviewed the three critical factors that influence the success of the MLSDLC process—namely, people, technologies, and processes.

By focusing on the *people* success factor, you also saw how a cross-functional team works together during the planning and design phases of the MLSDLC, each providing codified *technology* artifacts that meet the business objectives and shape the overall design of the solution.

However, we are not done yet! In the next chapter, we'll continue from where we've left off, with the platform team piecing the various artifacts into an E2E CI/CD/CT pipeline, thus automating the MLSDLC process.

11
Continuous Integration, Deployment, and Training for the MLSDLC

If you review some of the *Architecture Best Practices for Machine Learning* content, namely the *Build a Secure Enterprise Machine Learning Platform on AWS* whitepaper, and even the SageMaker documentation on MLOps, you will notice that among the various challenges of automating an application, they all call out the need to have a **cross-functional team**.

So, why is a cross-functional, agile team so important for automated ML on AWS?

AWS provides numerous ML-related *technologies* that often overlap in terms of their features to provide their customers with choice and flexibility. Furthermore, the industry provides many tried and tested *process* guidelines, such as CI/CD, to automate this process. However, neither AWS nor the industry can influence the organizational structure or application development culture of a company. Any changes need to happen within, and done by, the organization.

In *Chapter 10, An Introduction to the Machine Learning Software Development Life Cycle (MLSDLC)*, we focused on how a cross-functional team, made up of data scientists, ML engineers, and platform, application, data, and security experts all contribute to successfully implementing an automated MLSDLC process. By using a practical example of the *ACME web application*, you learned how these various personas interacted with each other, as well as why their domain expertise and artifact contributions are so important to the success of the project.

In this chapter, we are going to focus on automating the MLSDLC process to learn how the various artifacts we created in *Chapter 10, An Introduction to the Machine Learning Software Development Life Cycle (MLSDLC)*, map to each stage of the process.

To accomplish this, we will cover the following topics:

- Codifying the continuous integration stage
- Managing the continuous deployment stage
- Managing continuous training

By the end of this chapter, you will have completed an automated, end-to-end MLSDLC process that deploys the *ACME website*, along with the Age Calculator model, to production. This will provide you with the necessary framework to continually automate the process whenever any code changes are made or any new data is added.

Technical requirements

For this chapter, you will need the following:

- A web browser. (For the best experience, it is recommended that you use either Chrome or Firefox.)
- Access to the AWS account that you've been using throughout this book.
- Access to the Cloud9 development environment you've been using throughout this book.

- A reference to the usage limits of the AWS Free Tier to avoid unnecessary costs.

- The source code examples for this chapter, which are provided in this book's GitHub repository (`https://github.com/PacktPublishing/Automated-Machine-Learning-on-AWS/tree/main/Chapter11`).

Codifying the continuous integration stage

In this section, we are going to pick up where we left off in *Chapter 10, An Introduction to the Machine Learning Software Development Life Cycle (MLSDLC)*. We concluded the previous chapter with the various teams committing their artifacts to the source code repository. So, before a security review can take place, the team that plays the central role of integrating these artifacts into the overall solution, known as the Platform Team, takes the reins. At a high level, the following diagram shows how central a role the **Platform Team** plays in our scenario:

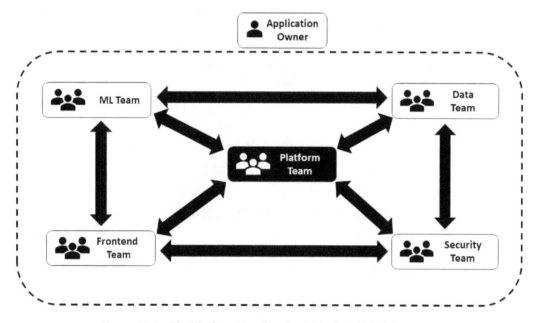

Figure 11.1 – The Platform Team's role within the MLSDLC process

As you can see, since the Platform Team sits in the *middle* of the cross-functional team, they are responsible for *gluing* all the solution components together. Once all the pieces have been *glued* together, the Platform Team is then responsible for verifying that these components function well together, as per the business use case. For example, the Platform Team would verify that a web application user can enter Abalone attribute data into the web UI and have this data sent as inference request data to the ML model, where the ML model returns a valid response to the user.

So, *how does the Platform Team integrate the various pieces together?* More importantly, *how does the Platform Team do this in a continuous and automated fashion?*

The best way to answer these questions is to practically showcase the typical tasks that are performed by the Platform Team as they build their integration artifacts.

Building the integration artifacts

To test whether all the pieces fit together, the Platform Team creates a mock-up of the production solution in a **test** or **Quality Assurance** (**QA**) environment. They then perform **functionality tests**, also called **system tests**, on the solution to ensure that the entire system works the way it's supposed to. Furthermore, to automate this process, the Platform Team codifies the solution as a **CDK construct**.

To build out this construct as the Platform Team, we are going to continue using the AWS Cloud9 IDE we used in *Chapter 10, An Introduction to the Machine Learning Software Development Life Cycle (MLSDLC)*. Follow these steps:

1. Log into the same AWS account you've been using throughout this book and open the AWS Cloud9 console (https://console.aws.amazon.com/cloud9).

2. In the **Your environments** section, click the **Open IDE** button for the **MLOps-IDE** development environment.

3. Using the Terminal window within the Cloud9 workspace, run the following commands to copy the pre-built stack construct from this book's GitHub repository into the stack.py folder:

```
$ cd ~/environment/acme-web-application
$ cp ~/environment/src/Chapter11/Files/cdk/test_
application_stack.py acme_web_application/stacks/
```

4. Using the left-hand navigation panel, double-click on the test_application_stack.py file to start reviewing it.

5. Now that the `test_applciation_stack.py` file is open, we can review the most important AWS resources that have been created by the stack construct. Besides loading the necessary CDK Python libraries and instantiating the `TestApplicationStack()` class, the first variable we must declare is `endpoint_name`. This is the name we will give to the SageMaker Hosted Endpoint, which is hosting our trained model.

6. Next, we must define an IAM role called `sagemaker_test_role`. This role will be used by SageMaker to access the Model Registry, where the production-grade model is stored.

7. The next variable we must define is the model itself. Here, we must instantiate the SageMaker model using the `CfnModel()` class of the SageMaker CDK module. We must also define an `AwsCustomResource()` to make an API call to the SSM service and retrieve the parameter that points to the location of the trained model within the Model Registry.

8. Now that we have defined the model, we need to allocate the compute resources that are required to host the model. This is done by instantiating the `endpoint_config` variable using the `CfnEndpointConfig()` class. Since this is for the test environment, we don't need to provide scalable compute resources – we just need to provide the bare minimum compute instances that are necessary to test the model. This is why we specified an `ml.t2.large` instance type for the test environment.

9. With both `model` and `endpoint_config` in place, we can instantiate `endpoint` using the `CfnEndpoint()` class, thus completing the model deployment part of the test environment.

10. The next component of the test environment is to create the back-service for the website's *Contact form* and the *Age Calculator form*. The Platform Team provides this backend functionality as a **RESTful API**, using the AWS API gateway service, by declaring the `api` variable as an `HttpApi()` gateway class. The team also distributes the static HTML components as part of a **Content Delivery Network (CDN)** using the `CloudFrontWebDistribution()` class.

> **Note**
>
> Since the CloudFront distribution is only used to test the various artifact integrations, we specify the distribution price class as `PRICE_CLASS_100`. This means that the static website content will only be distributed to edges in North America, South Africa, and the Middle East. By not using the full global distribution capabilities of CloudFront, we can minimize costs for testing. To learn more about CloudFront distribution classes and edge locations, you can view the pricing documentation (`https://aws.amazon.com/cloudfront/pricing/`).

11. Once the website's content has been uploaded to S3 and distributed through CloudFront, we can create routes for the *Contact form* and the *Age Calculator form* that will point to a **Lambda function** to process these requests. The `formHandler` Lambda takes the website API requests and handles them based on the `requests` path. For example, if the `formHandler` Lambda receives an API POST request from the `/api/predict` path, it will send the request payload to the SageMaker hosted model for inference. Then, it will take the inference response from the hosted model and send it back to the website.

12. Lastly, we must create two outputs using the CDK's `CfnOutput()` module. The first output is called `self.cdn_output` and contains `cdn.domain_name` as its value. This will allow us to capture the website URL.

13. The second output is called `self.api_output` and provides `api.url` as a value, essentially providing the URL for the form API.

We will be using these outputs in the next section to build the test artifacts.

Building the test artifacts

To test the application, we need to put ourselves in the shoes of the application user and learn how they may interact with the functionality that's provided within the web application. Since our example website only consists of an *HTML page*, a *Contact form*, and the *Age Calculator prediction form*, to test the overall functionality of the system, we must confirm that these components do what they are supposed to.

Follow these steps to create the necessary tests:

1. Using the Cloud9 workspace's Terminal window, run the following commands to copy the pre-built testing scripts into the `acme-web-application` folder:

```
$ cd ~/environment/acme-web-application
$ rm -rf tests
$ cp -R ~/environment/src/Chapter11/Files/tests .
```

2. Using the left-hand navigation panel of the Cloud9 environment, expand the `tests` folder, and then double-click on the `system_test.py` file to review the test code.

If you look at the test code, you will see that we use the Python `requests` library to simulate users making website requests. The *first test* focuses on the website itself by verifying that we get the appropriate status code back from the web server and that the delivered content is HTML code. In essence, this test simulates that the website is running and that it's accessible.

The *second test* focuses on the backend RESTful API. In this test, we send sample Abalone attribute data to the backend API, which, in turn, sends this to the hosted production model for inference. Then, we verify that we received the appropriate status code in return, along with an HTML response for the predicted Abalone age. In essence, this test simulates the user experience for the Age Calculator.

The *last test* simulates an incorrect call to the backend API to ensure that the API responds with the correct error messages. This test is not always necessary, but testing that the application responds with the correct errors ensures that when errors occur, they can be debugged correctly since we know that the application is reporting any errors correctly.

Now that we have scripted some basic functionality tests for the system, we can build out the production environment.

Building the production artifacts

Now that we've done the necessary, we have a fair idea that the tested system artifacts will work in production. So, to create the production environment, we must copy the application constructs we used in the test or QA environment. Follow these steps:

1. Using the Cloud9 workspace's Terminal window, run the following command to copy the pre-built production artifacts from this book's GitHub repository:

```
$ cd ~/environment/acme-web-application
$ cp ~/environment/src/Chapter11/Files/cdk/production_
application_stack.py acme_web_application/stacks/
```

2. Using the left-hand navigation panel, double-click on the `production_application_stack.py` file to review it.

If you compare the production application construct to the test application construct, you will notice that there are a few additional components. First, we created a new *S3 bucket* to store all the production application logs. This bucket will store all the logs for website access, as well as record the inference logs from the production model. You may recall from *Chapter 10, An Introduction to the Machine Learning Software Development Life Cycle (MLSDLC)*, when we discussed security, that it's a good practice to log all system activity. From the perspective of the production ML model, we log the inference request, as well as the inference response data, to gauge how well the model is performing.

When defining `endpoint_config`, you will see that, for the production environment, we use right-size production compute resources to host the model by using `ml.c5.large` instances. We also specify a minimum of 2 instances so that we can leverage the high availability (multiple AWS Availability Zones) features of the SageMaker Hosted Model. Additionally, we turn on inference via `data_capture_config` to log all inference request data and all response inference responses to the logging bucket.

Another new component we've added to the production construct is the `createBaseline` Lambda function. Since the production construct is deploying the production-grade ML model, we want to capture the statistical analysis of its expected performance. This way, by referencing the captured inference responses, we can monitor the model for quality drift. To this end, we defined the `baseline_creator` variable for the Lambda function, and then triggered the Lambda execution as a `CustomResource()`.

Finally, we added **endpoint auto-scaling**. This is the ability for the hosted model to be able to scale out and handle any increase in inference requests. We did this by defining the `scaling_target` variable and providing the policy, which specifies how the endpoint scales. For our production environment, we are going to start scaling when each `ml.c5.large` instance receives more than `750` requests per second over 15 minutes.

Within both the test and production constructs, we've instantiated the `formHandler` and `createBaseline` Lambda function. Both these variables refer to the code artifacts that comprise these functions. So, before we can close out the test and production CDK constructs, we need to update the source code respiratory with the pre-built Lambda artifacts to ensure that the constructs don't fail when we deploy them. Follow these steps to do so:

1. Using the Cloud9 workspace's Terminal window, run the following commands to copy the `formHandler` and `createBaseline` Lambda code into the cloned repository:

```
$ cd ~/environment/acme-web-application
$ cp -R ~/environment/src/Chapter11/Files/lambda/ .
```

With that, we have created all the necessary artifacts for the integration phase of the pipeline. Now, we must create the automation components for continuous integration by adding these components to the CDK Pipeline.

Automating the continuous integration process

In *Chapter 10, An Introduction to the Machine Learning Software Development Life Cycle (MLSDLC)*, we created a **skeleton CDK Pipeline**. This is referred to as a skeleton pipeline as we simply defined the pipeline construct, without providing the body or stages of the pipeline. So, now that most of the stack constructs, Lambda function, scripts, tests, and static HTML artifacts have been added to the repository, we can put them all together to create an automated CI/CD pipeline body. Follow these steps:

1. In your Cloud9 Terminal windows, run the following commands to update the `acme_pipeline_stack.py` construct:

```
$ cd ~/environment/acme-web-application
$ cp ~/environment/src/Chapter11/Files/cdk/acme_pipeline_
stack.py acme_web_application/
```

2. From the left-hand navigation panel, double-click the `acme_pipeline_stack.py` file for review.

If you compare the new `acme_pipeline_stack.py` file with the one we created in *Chapter 10, An Introduction to the Machine Learning Software Development Life Cycle (MLSDLC)*, you will see that there are some significant changes. First, we have imported the `cdk.Construct` classes from all the CDK stacks in the `stacks` folder. We've also defined a `cdk.Stage` class for each of the import stack constructs. For example, if you refer to the following excerpt, you will see that we imported the `MLWorkflowStack` class from the `ml_workflow_stack.py` file, which can be found in the `stacks` folder:

```
...
from .stacks.ml_workflow_stack import MLWorkflowStack
...
```

Then, we instantiate a class of this stack, called `MLWorklowStage()`, as a CDK Pipeline stage construct. We also supply the various parameters that are required to instantiate the stack as a pipeline stage and define the specific stack outputs (in this case, the ARN of the state machine):

```
...
class MLWorkflowStage(cdk.Stage):
    def __init__(self, scope: cdk.Construct, id: str, *, group_
name: str, threshold: float, data_bucket_name: str, feature_
group_name: str, **kwargs):
        super().__init__(scope, id, **kwargs)
        ml_workflow_stack = MLWorkflowStack(
```

```
        self,
        "MLWorkflowStack",
        group_name=group_name,
        threshold=threshold,
        data_bucket_name=data_bucket_name,
        feature_group_name=feature_group_name
    )
    self.sfn_arn = ml_workflow_stack.sfn_output
...
```

By instantiating all the CDK stacks as individual stage constructs, we are essentially defining each construct as a sequential part of the pipeline body. For example, if you scroll down to where we've defined the `PipelineStack()` class, you will see that the `MLWorkflow()` stage construct has been defined by the `ml_workflow_stage` variable. The `ml_workflow_stage` variable is, in turn, added to the pipeline body using the `add_stage()` **method** of the CDK Pipelines module.

> **Note**
>
> For more details on the `Stage()` construct and how to incorporate it into CDK Pipelines, please refer to the following AWS blog on CDK Pipelines: `https://aws.amazon.com/blogs/developer/cdk-pipelines-continuous-delivery-for-aws-cdk-applications/`. Keep in mind that this blog is based on the preview version of the CDK Pipelines module. In July 2021, AWS released CDK Pipelines as **generally available (GA)**. To review the differences between the preview and GA versions, you can refer to the API documentation (`https://github.com/aws/aws-cdk/blob/master/packages/%40aws-cdk/pipelines/ORIGINAL_API.md`).

Additionally, you will see that when we add the `ml_workflow_stage` and `test_stage` variables to the pipeline, we also define a `post` parameter. By using this parameter, we can define additional stage actions, or stage steps, that are executed after the stage construct has been deployed. In the case of the `ml_workflow_stage` variable, we instantiate an instance of the `CodeBuildStep()` class module to run a Python file called `invoke.py`. This script takes the ARN of the Step Functions state machine (deployed in the `ml_workflow_stage` construct) and starts executing the workflow. Alternately, in the case of `test_stage`, we instantiate the `ShellStep()` class module to run the `system_test.py` file, which tests the functionality of the application.

> **Note**
>
> The reason we use `CodeBuildStep()` for `ml_workflow_stage`
> and `ShellStep()` for `test_stage` is so that we can use the `role_`
> `policy_statements` parameter to supply the necessary IAM permissions
> to start and monitor the Step Functions state machine execution.

The last change you will see is that the pipeline's `self_mutation` parameter is now set
to `True`. This means that we are going to enable the pipeline's capability to dynamically
adapt (self-mutate) to any code changes. For example, if you open the CodePipeline
management console (`https://console.aws.amazon.com/codesuite/`
`codepipeline/`) for your region and click on **ACME -WebApp-Pipeline**, you will see
that the current structure of the pipeline only has two stages:

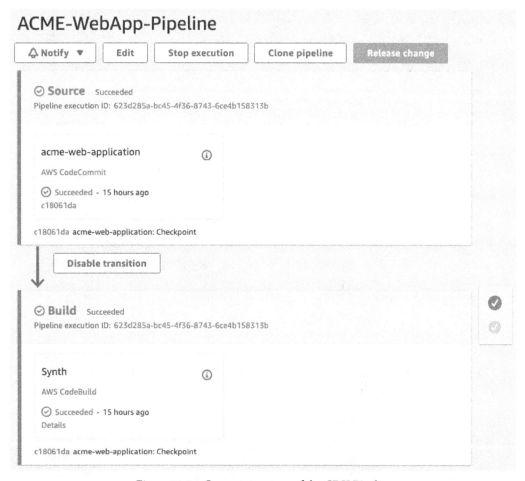

Figure 11.2 – Current structure of the CDK Pipeline

As we've been committing artifact updates to the source code repository, these changes have been triggering a pipeline execution. However, since the `self_mutation` parameter is currently set to `False`, adding stack and stage code constructs hasn't modified the pipeline structure.

Now, the Platform Team must finalize the CDK project to enable self-mutation. Follow these steps:

1. To finalize the CDK project, go to the Terminal windows in your Cloud9 workspace and run the following command to get the name of the SageMaker Feature Group:

```
$ aws sagemaker list-feature-groups --name-contains
abalone
```

> **Note**
>
> You will recall that this Feature Group is the Feature Group we created in the *Creating the SageMaker Feature Store* section of *Chapter 10, An Introduction to the Machine Learning Software Development Life Cycle (MLSDLC)*.

2. Copy the value for the `FeatureGroupName` key from the output.

3. Using the left-hand navigation panel of the Cloud9 workspace, expand the `acme-web-application` folder and double-click on the `app.py` file to start editing it.

4. Replace the `PLACEHOLDER` variable assignment with the output from the command you ran in *Step 1*, as shown in the following code snippet:

```
...
MODEL_GROUP = f"{MODEL.capitalize()}PackageGroup"
FEATURE_GROUP = "<Add the name of the SageMaker Feature
Group>"
CODECOMMIT_REPOSITORY = "acme-web-application"
...
```

5. Save and close the `app.py` file.

6. Using the following commands in your Cloud9 Terminal windows, add the final pipeline artifact – the `invoke.py` file – and commit the changes to the repository:

```
$ cd ~/environment/acme-web-application/
$ cp ~/environment/src/Chapter11/Files/scripts/invoke.py
scripts/
$ git add -A
```

```
$ git commit -m "Finalized CDK application"
$ git push
```

7. Since we've updated the CDK application, run the following commands to redeploy the application:

```
$ cdk deploy
```

Congratulations! You have just codified an automated ML-based application. However, we are still not done. The next step is to monitor the automated process to confirm that what we've created gets deployed into production and meets the functional requirements of the business use case. We'll be focusing on this task in the next section as we manage the continuous deployment of the codified solution.

Managing the continuous deployment stage

So far, we have focused primarily on the people that are involved in planning, designing, and codifying the solution. However, you will recall from *Chapter 10, An Introduction to the Machine Learning Software Development Life Cycle (MLSDLC)*, that outside of these *people*, two other factors influence the success of an MLSDLC implementation – the *technology* and the *process*. In this section, we are going to focus on the MLSDLC process itself. Since we have already codified the process using the self-mutating CDK Pipeline, all we need to do is manage the deployment to completion. To recap, let's review where we are in this process:

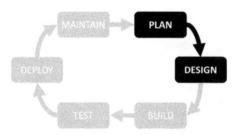

Figure 11.3 – The plan and design phases of the MLSDLC process

Here, you can see that we have already completed the **plan** and **design** phases of the MLSDLC process. As a cross-functional team, we have reviewed the business objectives and requirements for the *ACME web application*. Using the CDK, the various teams have codified their contributions to the design of the application. Now that the design has been deployed, we can move on to the next phase of the automated MLSDLC process – the **build** phase.

Reviewing the build phase

To review the build process, open the CodePipeline (`https://console.aws.amazon.com/codesuite/codepipeline/pipelines/`) management console for your current AWS region; you will see **ACME -WebApp-Pipeline**. Upon opening the pipeline, you will immediately see that the pipeline has self-mutated to incorporate the stages we've defined. Scrolling down the pipeline will reveal the **Assets** stage, as shown in the following screenshot:

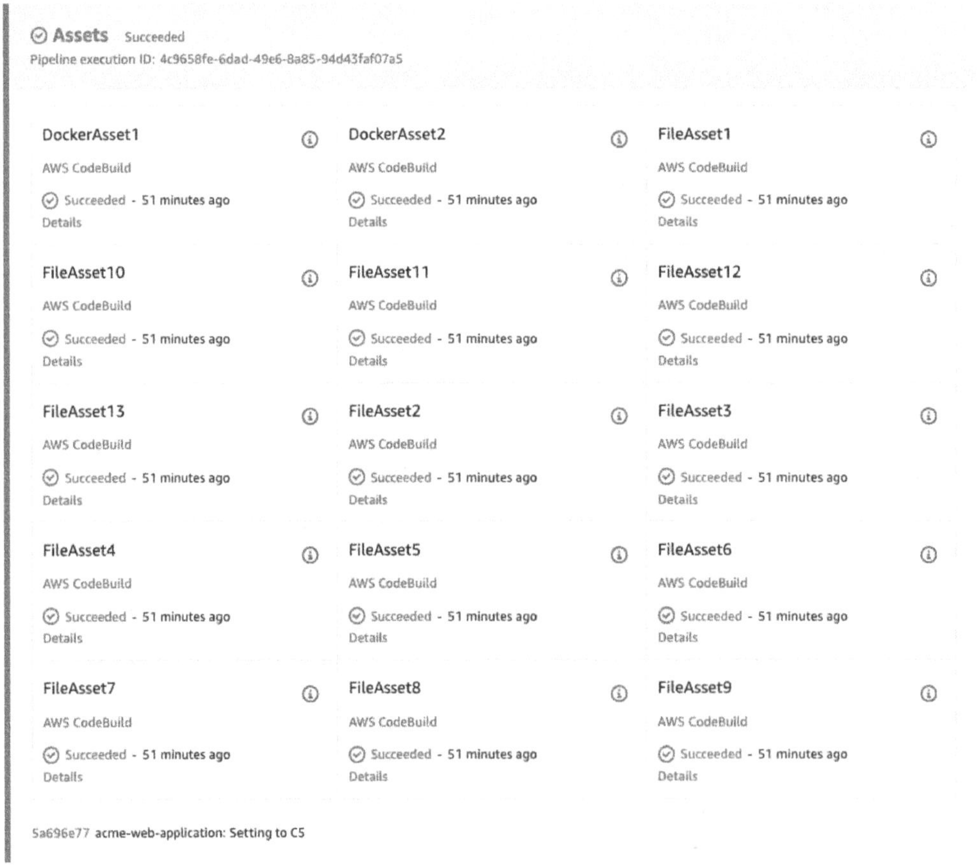

Figure 11.4 – The Assets stage

The **Assets** stage is the first part of the MLSDLC build phase, where the ML container image, the various Lambda function, and the static HTML web content are built. As you can see, we don't need to create a dedicated build stage to create these assets; the CDK Pipeline does this automatically. However, the Build process isn't completed once these pipeline assets have been created.

For the overall MLSDLC process to execute successfully, the build phase also requires a production-grade ML model. So, as shown in the following screenshot, scroll further down the pipeline to reveal the second part of the build process – the **Build-MLWorkflow** stage:

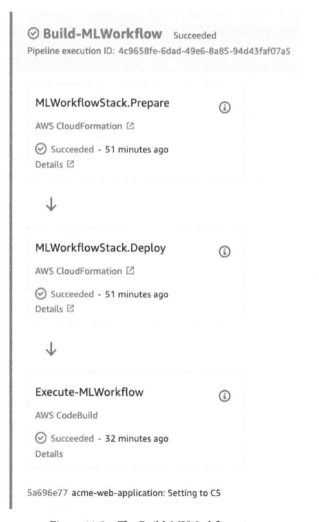

Figure 11.5 – The Build-MLWorkflow stage

As you can see, three separate actions make up the **Build-MLWorkflow** stage. These are the **Prepare, Deploy,** and **Execute** actions. The **Prepare** action creates a CloudFormation changeset to review the AWS resources that are being deployed by the stack, thus guaranteeing that any proposed changes don't impact existing, critical AWS resources. This is essentially a built-in integrity or integration test for the proposed resources within the context of continuous integration, where existing stacks are being automatically updated with pipeline changes. Since this is the first time the ML workflow is being created, the **Prepare** stage proceeds to the **Deploy** stage, where the stack is deployed using AWS CloudFormation.

Once the stack has been created, the **Execute-MLWorkflow** action is triggered. It's at this stage that the `invoke.py` script is run. Recall that the `invoke.py` script creates an execution of the Step Functions state machine. This state machine, in turn, trains a production-grade ML model.

> **Note**
> If you click on the **Details** link for the **Execute-MLWorkflow** action, you will be automatically redirected to the CodeBuild management console, whereby you can see the output from the `invoke.py` script within **Build logs**.

If you open the Step Functions console (`https://console.aws.amazon.com/states/home`) and click on the state machine name that starts with **MLWorkflow...,** you will see the list of **Executions**. Clicking on this reveals the current state of the workflow. Once the workflow has been completed, the execution graph should look as follows:

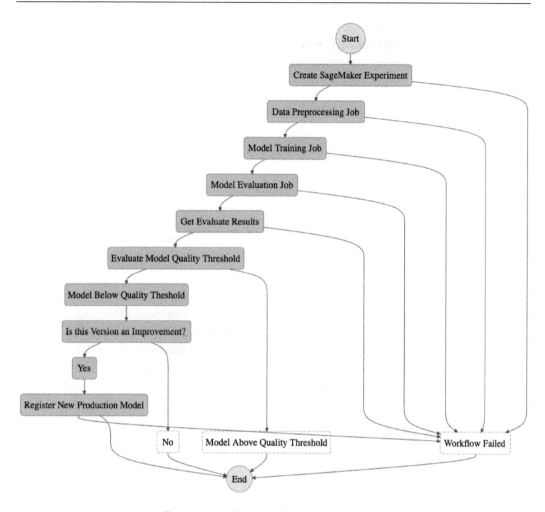

Figure 11.6 – State machine execution graph

As you can see, this is the first time the workflow has been executed. If the model's performance is below the threshold, it is added as a model package to the SageMaker Model Registry. The following screenshot shows an example of the model version metrics within the registry:

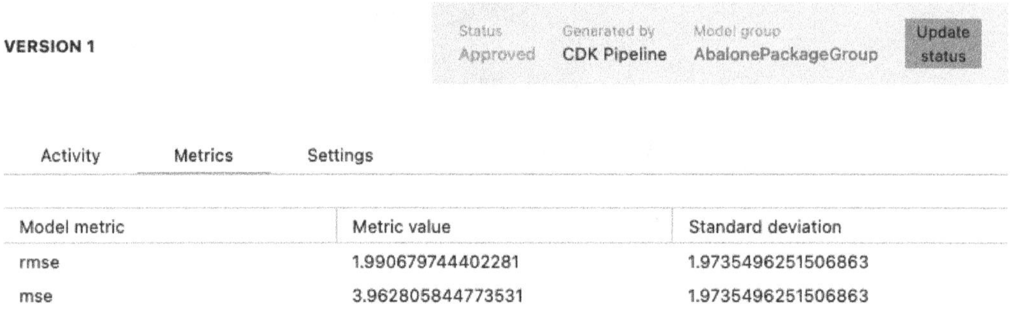

VERSION 1

	Status	Generated by	Model group	Update status
	Approved	CDK Pipeline	AbalonePackageGroup	

Activity   Metrics   Settings

Model metric	Metric value	Standard deviation
rmse	1.990679744402281	1.9735496251506863
mse	3.962805844773531	1.9735496251506863

Figure 11.7 – Model version metrics

As you can see, using the SageMaker Studio UI, the ML Team can track the lineage of the various models that have been produced by the workflow. Since we also enabled experiment tracking, the data processing, model training, and model evaluation trials are also available to the ML Team for assessment in the SageMaker Studio UI. The following screenshot shows an example of the training experiment that was produced by the workflow:

Experiment: AbaloneExperiments

Trial: Abalone-3057f2f9-9693-4136-ab4a-c4cc64dbe83f

Trial Component Created: 5 hours ago

Trial Component Status: Completed

Training job detailed status: Completed

Trial components: Training ▼

Charts   Metrics   Parameters   Artifacts   AWS settings   Debugger   Model explainability   Bias report   Trial Mappings

4 CHARTS   Add chart

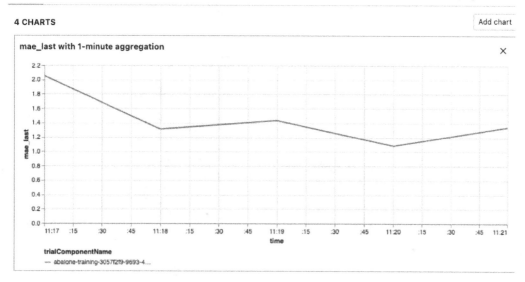

Figure 11.8 – Training experiment details

> **Note**
>
> For more information on how to compare SageMaker experiments and trials using SageMaker Studio, please refer to the following SageMaker documentation: `https://docs.aws.amazon.com/sagemaker/latest/dg/experiments-view-compare.html`.

Since executing the ML workflow is the final action within the **Build-MLWorkflow** stage, we have completed the build phase of the MLSDLC process. At this stage, the pipeline automatically moves on to the **test** phase.

Reviewing the test phase

Once the various pipeline assets and the production-grade ML model have been built, we must move on to the test phase, as shown in the following diagram:

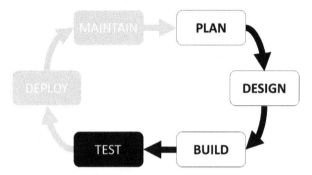

Figure 11.9 – The test phase of the MLSDLC process

During the test phase, our pipeline deploys a pseudo-production version of the solution into a test or QA environment. If we review this stage of the pipeline in the CodePipeline console, you will see that there are also three stage actions to this **Test-Deployment** stage. The following screenshot shows an example of the **Test-Deployment** stage:

Figure 11.10 – The Test-Deployment stage

As you can see, the **Test-Deployment** stage also has **Prepare** and **Deploy** stage actions. Since these actions are pre-built by the CDK Pipeline, they perform the same activities as the related stage actions within the **Build-MLWorkflow** stage, except that instead of deploying the ML workflow assets, a pseudo-production solution is deployed for system testing. Once the environment has been deployed through CloudFormation, the **System-Tests** stage action runs the system_test.py file to perform these three system tests.

> **Tip**
>
> Once the **System-Test** stage action has been completed, you can delete the **Test-Deployment-TestApplicationStack** CloudFormation Stack via the CloudFormation console (`https://console.aws.amazon.com/cloudformation/home`). We don't have any further requirements for these resources and we don't wish to incur any further AWS usage costs from them being idle.

As you may recall from the previous section, these three tests simulate the user experience with the solution by accessing the website and sending ML inference requests to the Age Calculator model. By clicking on the **Details** link for the **System-Tests** action, you will see the CodeBuild **Build logs** output from running the system tests.

> **Note**
>
> If the system test script should fail for whatever reason, the **System-Test** action and, consequently, the **Test-Deployment** stage will fail. Having any of these tests fail doesn't necessarily mean that the MLSDLC process will fail. The whole point of automating the MLSDLC process, especially automating the system tests, is to verify that once the solution is eventually deployed into production, we can be confident in its functionality. So, if the tests fail, we can provide debugging feedback to the supporting team, who can, in turn, resolve the issue and re-execute the pipeline.

Now that the **Test-Deployment** stage of the pipeline is complete and we have a tested solution, we are ready to deploy the solution into production. This is known as the **deploy** phase of the MLSDLC.

Reviewing the deploy and maintain phases

Once all the system tests have been run on the pseudo-production solution, we should be confident that the production version is ready for our users. As shown in the following diagram, we are ready to finally deploy the solution to production. Once deployed, we can manage and maintain it:

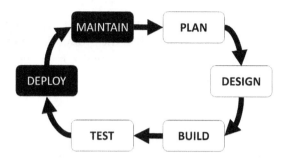

Figure 11.11 – The deploy and maintain phases of the MLSDLC process

From the standpoint of the CDK Pipeline, the stage that's responsible for production deployments is the same as the test deployment stage in that this stage also has the **Prepare** and **Deploy** stage actions, but no system testing stage actions. The following screenshot shows an example of only these two actions being performed within the **Production-Deployment** stage:

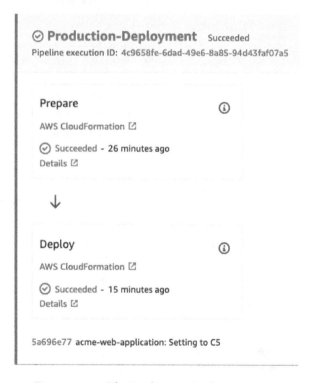

Figure 11.12 – The Production-Deployment stage

While the pipeline stages may be similar, the solution that's being deployed as a CloudFormation stack is somewhat different. First, the production stack deploys optimal AWS resources that are better suited to a production environment. For example, the production stack uses optimized C5 compute resources to host the model and implements additional elasticity in that these compute resources can scale out, as well as scale back in, depending on user demand.

Moreover, since the **Maintain** phase of the MLSDLC is an operational activity, this means it can't be automated easily unless you apply some type of **Artificial Intelligence Operations** (**AIOps**) methodology. In this example, however, we do facilitate automated maintenance in the production stack. For example, you will recall from the previous section that when we codified the `production_application_stack.py` file, we enabled `logging_config` for the `CloudFrontDistribution()` class. This enables easier maintenance of the solution once it's deployed since we store all the web transaction logs in S3. This gives the operations teams the ability to see what's going in within the stack and use this information for troubleshooting and debugging purposes.

Apart from this, you will recall that, in the `production_application_stack.py` file, we also created the `createBaseline` Lambda function and invoked it using the `CustomResource` CDK module. In the following code snippet, which has been taken from the Lambda function's `index.py` file, you can see that this function runs a SageMaker Processing Job to perform statistical analysis of the testing data. It does this using the `sagemaker-model-monitor-analyze` container, which is provided by AWS, to baseline the expected performance of the trained model:

```python
...
image_map = {
    "us-east-1": "156813124566.dkr.ecr.us-east-1.amazonaws.com/
sagemaker-model-monitor-analyzer",
...
        logger.info(f'Creating Basline Suggestion Job:
{request["ProcessingJobName"]}')
        try:
            response = sm.create_processing_job(**request)
            return {
                "PhysicalResourceId":
response["ProcessingJobArn"],
                "Data": {
                    "ProcessingJobName":
request["ProcessingJobName"],
                    "BaselineResultsUri": f"s3://{logs_bucket}/
```

```
baselining/results"
                    }
            }
. . .
```

By combining this statistical baseline analysis with the captured inference response data from the production model, the operations teams can detect if the model is drifting from its intended purpose.

Furthermore, by facilitating both the data capture and baseline data, the operations team can automate the drift detection process by implementing SageMaker Model Monitor (https://docs.aws.amazon.com/sagemaker/latest/dg/model-monitor. html). Model Monitor will use these data sources to detect various kinds of model drift automatically, on a predefined schedule.

Note

We have not implemented the automated model monitoring capabilities with this example since multiple types of built-in drift detection capabilities are provided by the Model Monitor module. You can review the documentation to determine whether data quality (https://docs.aws.amazon.com/ sagemaker/latest/dg/model-monitor-data-quality. html), model quality (https://docs.aws.amazon.com/ sagemaker/latest/dg/model-monitor-model-quality. html), bias drift (https://docs.aws.amazon.com/sagemaker/ latest/dg/clarify-model-monitor-bias-drift. html), or feature attribution drift (https://docs.aws.amazon. com/sagemaker/latest/dg/clarify-model-monitor- feature-attribution-drift.html) suits your production use case requirements.

Once the **Production-Deployment** stage of the pipeline is complete, we will see that we closed the loop and completed the MLSDLC process, as shown in the following diagram:

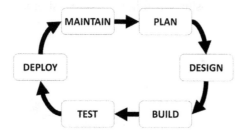

Figure 11.13 – Completed MLSDLC process

Let's look at what we've built by reviewing what our users may experience when using the *ACME web application* and the *Age Calculator* component.

Reviewing the application user experience

To review the production application, open the CloudFormation console (`https://console.aws.amazon.com/cloudformation/home`) and click the radio button next to **Production-Deployment-ProdApplicationStack** to open the stack. Click on the **Outputs** tab to view the stack outputs, as shown in the following screenshot:

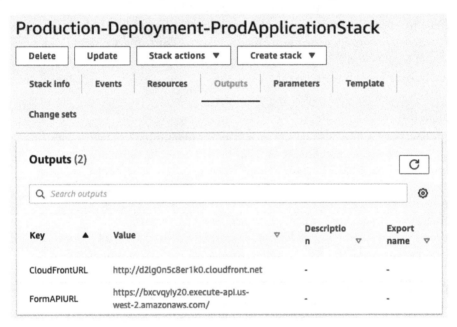

Figure 11.14 – CloudFormation stack outputs

As you can see, we have two stack outputs. The **FormAPIURL** output is the API gateway address that's used to process the *Age Calculator* inference requests, while the **CloudFrontURL** output points to the address of the website. Click on **Value** for **CloudFrontURL** to view the website. The following screenshot shows the ACME Fishing Logistics website:

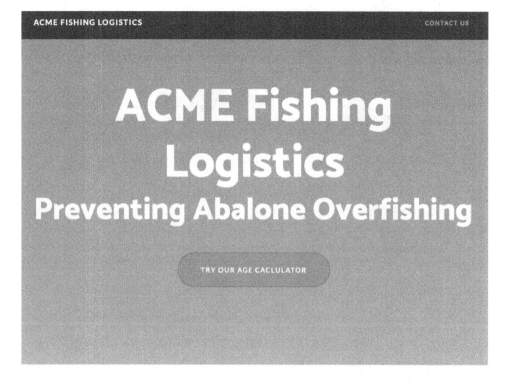

Who are we?

We are an organization thst strives to prevent the overfishing of the abalone sea snail. We do this by using cutting edge, predictive Artifical Intelligence (A.I.) technologies to help abalone fishermen determine wether or not their catch meets the age requiremnts for capture.

"Team work", from pxfuel, licensed under CC0 Public Domain

Figure 11.15 – ACME Fishing Logistics website

Now, let's try the *Age Calculator* component to see how a fisherman would be able to see the predicted age of his Abalone catch. The following screenshot shows the Age Calculator form that appears when a fisherman clicks the **TRY OUR AGE CALCULATOR** button:

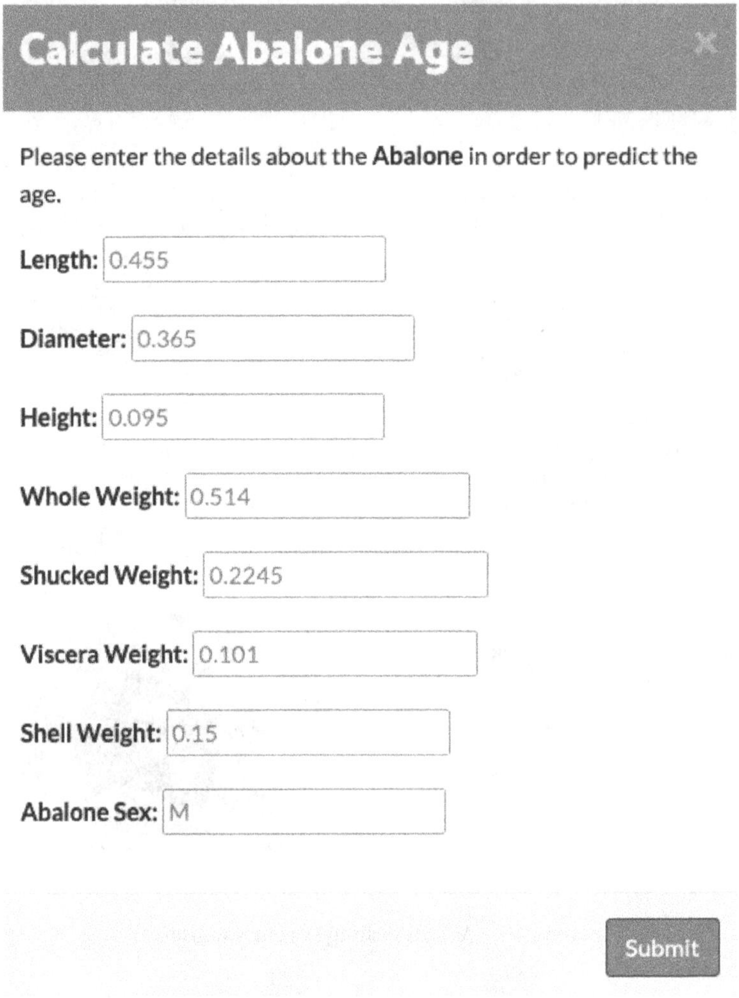

Calculate Abalone Age   ✕

Please enter the details about the **Abalone** in order to predict the age.

Length: 0.455

Diameter: 0.365

Height: 0.095

Whole Weight: 0.514

Shucked Weight: 0.2245

Viscera Weight: 0.101

Shell Weight: 0.15

Abalone Sex: M

Submit

Figure 11.16 – Age Calculator form

As you can see, the **Calculate Abalone Age** form provides various sample dimensions of the Abalone. Enter these sample dimensions and click the **Submit** button to see what the ML model predicts. The following screenshot shows an example response from the trained model:

Age Prediction

We've calcuated that the Abalone has **10** rings, and is therefore
approximately **11.5** years old.

Close

Figure 11.17 – Age Prediction

As you can see, based on the sample dimensions provided, the model predicts that
the Abalone has **10** rings. According to the UCI Machine Learning Repository for the
Abalone dataset (`https://archive.ics.uci.edu/ml/datasets/abalone`), the
value for the number of **Rings**, plus **1.5**, gives us the age in years. So, a fisherman can see
that the Abalone is **11.5** years old and thus determine whether the catch should be thrown
back or kept.

Congratulations! We now have a working web application and a built-in ML model for
our fisherman customers. We used an automated MLSDLC process to accomplish this
business objective.

However, you will recall that an MLSDLC process differs from a typical SDLC process
in that we are not only continuously automating the release of an ML-based application
when the business case or source code changes, but also when the training data changes.
Remember, an ML model is only as good as the data it's trained upon. So, *how do we
continuously automate the MLSDLC process when we have new data?*

In the next section, we will answer this question by exploring the concept of **continuous
training (CT)**.

Managing continuous training

In *Chapter 9, Building the ML Workflow Using Amazon Managed Workflow for Apache Airflow*, we learned how Airflow can be used to create a data-centric ML process and train the *Age Calculator* model on new Abalone survey data. In *Chapter 10, An Introduction to the Machine Learning Software Development Life Cycle (MLSDLC)*, we learned how the Data Team applied this technique to the *ACME web application* by codifying the **acme-data-workflow** Airflow DAG. The following diagram shows a graphical representation of the Airflow DAG:

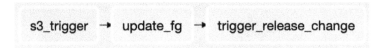

Figure 11.18 – Data Airflow DAG

As you can see, the Airflow DAG starts when new Abalone survey data is added to the S3 bucket. The survey data is then preprocessed to engineer the relevant training features; these features are then ingested into the Feature Store. Once the new data is ingested into the Feature Store, a release change of the MLSDLC process is triggered to automate the process of releasing a new changeset of the solution. Essentially, this creates a continuous training process.

Moreover, at the beginning of this chapter, we saw how the Platform Team incorporated this concept of continuous training into the CI/CD methodology by extending the CDK Pipeline to provision the necessary AWS recourses that manage and execute the **acme-data-workflow** DAG. For instance, if you open **ACME-WebApp-Pipeline** in the CodePipeline console, you will see the **Build-Data-Workflow** stage, as shown in the following screenshot:

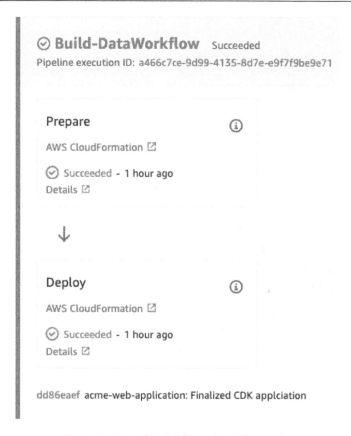

Figure 11.19 – The Build-DataWorkflow stage

As you can see, the **Build-DataWorkflow** stage has both a **Prepare** and a **Deploy** stage action, whereby a CloudFormation changeset is prepared and then deployed. The result of this deployment is an MWAA environment inside a VPC, plus the DAG, and its supporting assets uploaded to S3. Since this is the last stage of the CDK Pipeline, we have finally created a CI/CD/CT pipeline to incorporate continuous training into the MLSDLC process.

However, before we see the end-to-end MLSDLC process in its entirety, we need to simulate adding new Abalone survey data. We'll do this in the next section.

Creating new Abalone survey data

In *Chapter 9, Building the ML Workflow Using Amazon Managed Workflow for Apache Airflow*, we leveraged the `CTGAN` Python library to synthesize new Abalone data within a Jupyter Notebook. The following steps will walk you through reproducing the same process using SageMaker Studio and running the pre-built notebook in this book's GitHub repository (`https://github.com/PacktPublishing/Automated-Machine-Learning-on-AWS/tree/main/Chapter11/Notebook`):

1. Open the SageMaker management console (`https://console.aws.amazon.com/sagemaker/home`) and, in the left-hand panel, click the **Studio** link, under the **SageMaker Domain** section.

2. Once the **SageMaker Domain** dashboard opens, click the **Launch app** dropdown and select **Studio** from the list to open the Studio IDE.

3. Within the left-hand file panel, expand the `Notebooks` folder within the `Chapter11` folder of the cloned `src` folder.

> **Note**
>
> This book's GitHub repository files should have already been cloned into the Studio environment for you to use. If not, please refer to the *Creating a SageMaker Feature Store* section of *Chapter 10, An Introduction to the Machine Learning Software Development Life Cycle (MLSDLC)*.

4. Double-click on the `Simulating New Abalone Survey Data.ipynb` file to open the notebook.

5. From the **Kernel** menu, click the **Restart Kernel and Run all Cell...** option.

6. Once you've created the notebook, you can close the SageMaker Studio UI.

Now that we have simulated new Abalone survey data and uploaded the dataset to S3, we can review the continuous training process in action.

Reviewing the continuous training process

When the Data Team originally defined the Airflow DAG in the `continuous_training_pipeline.py` file, they used the `S3PrefixSensor()` provider to constantly check the S3 bucket for new data. So, now that we have simulated new Abalone survey data, the Airflow DAG should start running.

However, the DAG needs to be manually enabled for it to start running. To see the continuous training process in action and enable the DAG, follow these steps:

1. Open the MWAA console (`https://console.aws.amazon.com/mwaa/home`) and click the **Open Airflow UI** link for **acme-airflow-environment**.

2. Once the Airflow UI opens, toggle the **Pause/Unpause DAG** button next to the **acme-data-workflow** DAG to enable it. The DAG should automatically start.

3. Click on the **DAG** link to view the run. Once the DAG dashboard opens, click the **Graph View** button to see it represented as shown in *Figure 11.18*.

4. You can follow the DAG's progress and see the logs for each task by clicking on a specific task and clicking on the **Log** button.

5. Once each task has been run, the completed graph should look as follows:

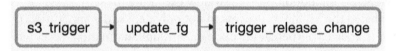

Figure 11.20 – Completed data workflow graph

As a result of the DAG run completing, you can reopen the CodePipeline console to see the **ACME-WebApp-Pipeline** restart. With that, you have just created a CI/CD/CT pipeline that continuously builds and deploys an ML-based application by automating the MLSDLC process on AWS.

Cleanup

To avoid unnecessary AWS usage costs, you can delete the resources that have been created by the CDK Pipeline by opening the CloudFormation console and then deleting the various stacks in the reverse order they were created. For example, select **Build-DataWorkflow-DataWorkflowStack** and then click the **Delete** button. Once this stack has been deleted, do the same for **Production-Deployment-ProdApplicationStack**.

> **Note**
> Depending on the stack that's being deleted, you may need to manually empty the S3 bucket for a particular stack before the stack can be deleted.

Continue doing this by going down the list of CloudFormation stacks until the **acme-web-application** stack has been deleted. That concludes this chapter.

Summary

In the final chapter of this book, you were introduced to the concept of MLSDLC and how this methodology can be used to create ML-based applications. Throughout the last two chapters, we have focused on two of the three primary success factors that are required to create an ML-based application – the people and the process.

By focusing on how a cross-functional team and an agile team cooperate, we learned how each team contributes their domain expertise to address both the business plan requirements and the solution design requirements of the MLSDLC.

The practical outcome of this exercise was a set of codified CDK stack constructs that, when *glued* together by the Platform Team, created a CI/CD/CT pipeline. This CI/CD/CT pipeline functioned and is the mechanism behind achieving MLSDLC methodology automation. For example, with each stage of the pipeline corresponding to a particular phase of the MLSDLC methodology, we saw how executing the CI/CD/CT pipeline inevitably automated the MLSDLC process to not only deploy the application into production but establish a perpetual life cycle of constant automation.

While these chapters did not specifically pay attention to the technology aspect of a successful MLSDLC methodology, it was evident how AWS technologies enabled the MLSDLC process.

So, by adding these technologies into the mix, in this chapter, we've successfully demonstrated an end-to-end example of automated ML on AWS.

Congratulations! You've made it to the end of this book. Now, you should have enough code references to insert some ML models and automate them on AWS.

Further reading

The following are some references to AWS content that highlight the importance of a cross-functional team as the key to successful ML automation on AWS:

- *Architecture Best Practices for Machine Learning*: `https://aws.amazon.com/architecture/machine-learning/`
- *Build a Secure Enterprise Machine Learning Platform on AWS*: `https://docs.aws.amazon.com/whitepapers/latest/build-secure-enterprise-ml-platform/build-secure-enterprise-ml-platform.html`
- *SageMaker MLOps Documentation*: `https://docs.aws.amazon.com/sagemaker/latest/dg/sagemaker-projects-why.html`

Index

C

X

Y

Subscribe to our online digital library for full access to over 7,000 books and videos, as well as industry leading tools to help you plan your personal development and advance your career. For more information, please visit our website.

Why subscribe?

- Spend less time learning and more time coding with practical eBooks and Videos from over 4,000 industry professionals

- Improve your learning with Skill Plans built especially for you

- Get a free eBook or video every month

- Fully searchable for easy access to vital information

- Copy and paste, print, and bookmark content

Did you know that Packt offers eBook versions of every book published, with PDF and ePub files available? You can upgrade to the eBook version at packt.com and as a print book customer, you are entitled to a discount on the eBook copy. Get in touch with us at customercare@packtpub.com for more details.

At www.packt.com, you can also read a collection of free technical articles, sign up for a range of free newsletters, and receive exclusive discounts and offers on Packt books and eBooks.

Other Books You May Enjoy

If you enjoyed this book, you may be interested in these other books by Packt:

Automated Machine Learning with Microsoft Azure

Dennis Michael Sawyers

ISBN: 978-1-80056-531-9

- Understand how to train classification, regression, and forecasting ML algorithms with Azure AutoML
- Prepare data for Azure AutoML to ensure smooth model training and deployment
- Adjust AutoML configuration settings to make your models as accurate as possible
- Determine when to use a batch-scoring solution versus a real-time scoring solution
- Productionalize your AutoML solution with Azure Machine Learning pipelines
- Create real-time scoring solutions with AutoML and Azure Kubernetes Service
- Discover how to quickly deliver value and earn business trust using AutoML
- Train a large number of AutoML models at once using the AzureML Python SDK

Learn Amazon SageMaker - Second Edition

Julien Simon

ISBN: 978-1-80181-795-0

- Become well-versed with data annotation and preparation techniques
- Use AutoML features to build and train machine learning models with AutoPilot
- Create models using built-in algorithms and frameworks and your own code
- Train computer vision and natural language processing (NLP) models using real-world examples
- Cover training techniques for scaling, model optimization, model debugging, and cost optimization
- Automate deployment tasks in a variety of configurations using SDK and several automation tools

Packt is searching for authors like you

If you're interested in becoming an author for Packt, please visit `authors. packtpub.com` and apply today. We have worked with thousands of developers and tech professionals, just like you, to help them share their insight with the global tech community. You can make a general application, apply for a specific hot topic that we are recruiting an author for, or submit your own idea.

Share Your Thoughts

Now you've finished *Automated Machine Learning on AWS*, we'd love to hear your thoughts! Scan the QR code below to go straight to the Amazon review page for this book and share your feedback or leave a review on the site that you purchased it from.

https://packt.link/r/1801811822

Your review is important to us and the tech community and will help us make sure we're delivering excellent quality content.

www.ingramcontent.com/pod-product-compliance
Lightning Source LLC
Chambersburg PA
CBHW081502050326
40690CB00015B/2898